Praise for *So You Want to Be a*

Three years ago, we moved from our small hobby farm in Virginia to the woods of Maine. While we considered ourselves ready, nothing could have prepared us for the rigors of heating our home with wood, maintaining acres of fields, gardening in chilly zone 5 or the threat of coyotes and bears to our livestock and pets, not to mention living more than 40 minutes from, well, anything. If only I had Kirsten's book back then. Well-written and engaging, this book relates a life-long Mainer's experience restoring her farmstead, sprinkling questions throughout that the burgeoning homestead would do well to ask themselves before diving in to the self-sufficient lifestyle.

—Lisa Steele, author of *Fresh Eggs Daily* and *101 Chicken Keeping Hacks*

The pragmatic and practical idealism of Kristin's book will bless many current and want to become homesteaders—you will walk away with a realistic look at the joys and trials, benefits and challenges that a homestead life offers, along with a great deal of "worked for wisdom" that they share along the way.

—John Moody, author of *The Frugal Homesteader*

In her book *So You Want to be a Modern Homesteader*? author Kirsten Lie-Nielsen offers a frank and beautifully honest framework for getting from where you are today to that place of pastoral self-reliance collectively known as homesteading. Is it going to be easy? No! Is it going to take planning and personal fortitude and resilience? Yes! Can you be a modern homesteader and still earn a living? Possibly! With a no-nonsense, yet delightful approach, Lie-Nielsen guides the reader through the thought exercises, property analyses, fundamental skillsets and virtually everything else you need to maximize your enjoyment and successes on any homesteading journey. The book is perfect for folks just beginning the journey, experienced old-timers like me and everyone in between because self-reliance is a journey, not a destination.

—Hank Will, Editorial Director, Ogden Publications

The most valuable aspect of Kirsten's book is her absolute honesty about the emotional aspects of homesteading, as well as her refusal to narrowly define a modern homesteader. I love that every chapter ends with a list of "Questions Before You Leap" for the reader to answer. This book is especially valuable for someone who is just starting to contemplate a homesteading lifestyle and wondering if they should jump in—and how far.

—Deborah Niemann, ThriftyHomesteader.com,
author of *Homegrown and Handmade* and *Raising Goats Naturally*

Kirsten Lie-Nielsen's new book brings a realism and humor to her stories and experiences of going "rural". The book proves a helpful guideline to some of the pitfalls that folks dreaming of the idyllic countryside might not expect, while still embracing all that a lifestyle in the country has to offer. A perfect starting place for the would-be homesteader, this engaging book is a fun take on all things homesteading.

—Diana Rodgers, The Sustainable Dish

So You Want to Be a Modern Homesteader?

So You Want to Be a Modern Homesteader?

All the Dirt on Living the Good Life

Kirsten Lie-Nielsen

Cover design by Diane McIntosh.
Cover images: Main image © Kirsten Lie-Nielsen;
wood background (bottom) ©iStock 637668596.
Interior images: chapter number icon © sodesignby;
chapter end icon © ERNEST / Adobe Stock.

Printed in Canada. First printing November 2018.

Any other inquiries can be directed by mail to:

New Society Publishers
P.O. Box 189, Gabriola Island, BC V0R 1X0, Canada
(250) 247-9737

LIBRARY AND ARCHIVES CANADA CATALOGUING IN PUBLICATION

Lie-Nielsen, Kirsten, 1990–, author
So you want to be a modern homesteader? : all the dirt on living the good life / Kirsten Lie-Nielsen.

Includes index.
Issued in print and electronic formats.
ISBN 978-0-86571-891-3 (softcover).—ISBN 978-1-55092-684-2 (PDF).—
ISBN 978-1-77142-280-2 (EPUB)

1. Country life. 2. Self-reliant living. 3. Sustainable living. I. Title.

GF78.L54 2018 640 C2018-905283-X
C2018-905284-8

Funded by the Government of Canada | Financé par le gouvernement du Canada | Canada

New Society Publishers' mission is to publish books that contribute in fundamental ways to building an ecologically sustainable and just society, and to do so with the least possible impact on the environment, in a manner that models this vision.

Contents

Introduction

As I wrote this book, I kept coming up against the same question over and over again: What exactly is a homesteader? The term is inexact and open to interpretation. People from all walks of life rally around this word. There are preppers and survivalists, looked at with skepticism by hippies and wellness mamas, and these groups might be warily judged by small family farmers. Yet all of them, in a pinch, would self-define as homesteaders.

It could be said that the only true homesteaders were those relocating under the 1862 Homestead Act. Signed by Abraham

Credit: Kate St. Cyr

Modern homesteading life.

Lincoln, this act gifted 160 acres of land to anyone willing to relocate to the western territories. That's when the term "homesteader" first started being used in the manner that we use it today, though the "back-to-the-land" movement is much older than that. It seems that over the history of civilization, there have always been people wanting to drop out of the mainstream to live a more agrarian lifestyle.

The most recent periods of rural migration in American history happened during the 1960s and 1970s when "back-to-the-landers" flooded into rural states, such as Maine, following the lead of proponents such as Helen and Scott Nearing. Before that, during the late 1800s, folks were settling out West, while back East transcendentalists such as Henry David Thoreau waxed lyrical about the simple life. But you can find "homesteaders" in the Roman era and throughout world history. For the most part when looking at history, I define a homesteader as someone like the Roman dictator Lucius Quinctius Cincinnatus, who preferred life on his small farm to governing the young Roman Empire. These are people who had a choice and preferred living off the land, rather than taking jobs in the city. Of course, there were many yeoman farmers and laborers throughout history, but to begin to define the term "homesteader," I would offer it is someone who makes a conscious choice to live close to the land.

While dictionary definitions and blog posts on homestead websites extoll what individual homesteaders do, self-reliance is always at the heart of this lifestyle choice. A goal for self-sufficiency is borne from a desire to control your own destiny, or, as Thoreau put it, "to live so sturdily and Spartan-like as to put to rout all that was not life." Negative experiences within society can make one feel more comfortable outside of the mainstream, or it can derive from a fear or rejection of a direction that culture seems to be heading, and a will to survive even if society collapses.

So my broad definition of a homesteader would be "someone who chooses to live self-reliantly."

At first, I worried about the content. Should I spend more time explaining the workings of a woodstove, and skim the topic of no-till gardens? On my homestead, all our winter heat comes from a woodstove, and I have yet to tend a no-till garden. In the end, I decided to offer readers my own hard-earned, practical knowledge rather than hearsay where ever possible.

I questioned myself: Am I really a homesteader? The more I thought about it, the more I wondered if anyone was really qualified. I know people who would self-identify as homesteaders who live completely off-grid, with no electricity whatsoever, no running water, but happily drive to the local store to purchase food for their family. I know people who live in comfortable houses with all the amenities, but grow all of the food for their family, make their skin lotions and soaps, even sew their own clothing. And then there is me, in between: no indoor plumbing and fully dependent on wood heat, but using electricity and growing as much of our food as possible, while still making weekly pilgrimages to the grocery store for what we cannot supply ourselves.

With such a broad definition, *someone who chooses to live self-reliantly*, it's clear to me there is no right or wrong way to homestead. It is about doing what is within your abilities in order to create a more self-sufficient lifestyle. For some, it's about making or producing items you would otherwise purchase. For others, it's about withdrawing from modern society and abandoning the scrutiny from unwanted authorities. While these particular homesteaders may not completely share the same mindsets and opinions, both check off "homesteading" as their lifestyle choice.

So maybe your self-reliance starts in the kitchen, or in the garden, or raising livestock for food and fiber, or maybe both. This book is directed at folks considering a shift in their lifestyle from city-based to country living. You might even take the information and apply it to an urban homestead as well. My intention is to offer tips and strategies, as well as personal observations, regarding what to expect when moving to a rural area and to illustrate the pitfalls and perks that you might not have anticipated. This book also provides a practical guide to managing some of the realities of homestead life, including the dusty and sometimes finicky woodstove.

It is meant neither to encourage nor discourage, but to be realistic and informative. Homesteading in the country has many harsh truths, and it requires a strong work ethic above all other qualifications. It also has brought more joy and satisfaction to my existence than any other lifestyle I have known, and I hope that if you embark on such a journey it will offer you the same delights.

CHAPTER 1

Preparing for Rural Life

In April 2016, my partner and I packed up our little hobby farm and moved to the country. We had kept half a dozen geese and a small garden in a little enclave of mid-coast Maine. The closest neighbors were one hundred feet away, and their front windows looked directly into our back windows. We had built a small coop and run for our birds and spent a summer erecting a stone wall around the garden for some semblance of privacy from the busy route that ran directly in front of our home.

Proud new owners of an old homestead.

Our sign that we needed to escape suburbia? After our geese escaped their run for the umpteenth time and spent an afternoon sunning themselves along the side of the road, we had a knock on the door from animal control. We were told that we'd be responsible for damages if someone hit one of the geese.

We had already been looking for a place that offered more space and privacy, and now we began to search in earnest.

It took us years to find the right place. We looked at vacant acres with thoughts to build, and seriously considered a house at the end of a dead-end road three hours from anywhere. Our budget was tight, and while our priority was land, we knew it would be some time before we had any extra funds for building, so we wanted a place with some structures for ourselves and our animals. And while we longed for privacy, we also wanted to be within driving distance of our families.

We finally found the place, and knew on the first drive-by that it was the home for us. Nestled on 93 acres an hour inland from our old location, our new home was a farmhouse that had been kept in the same family since it was built in the 1860s. The home had been abandoned for more than thirty years and had no electricity or running water, and the fields were covered in low brush and surrounded by encroaching forests. But the property did feature a structurally sound, massive Yankee-style barn, which was full to the brim with decades of farming refuse.

Located on a road memorably titled Hostile Valley, our farm would never be known to the locals as anything other than "the

"The old Whitaker place" six months after we started clearing the fields.

old Whitaker place." Once a thriving family farm, it had slowly lost its residents until it became a summer place and then a memory. The bones of both buildings were strong, and if you squinted, you could imagine the long, rolling fields as open acres once again. It was time to pick up our lives and make the move to a real rural lifestyle.

Having lived most of my life in a less populated state, Maine, and enjoyed the ideas of privacy and solitude, I thought I was prepared for this transition. In many ways, I was. After all, it was the fulfillment of a lifelong dream. But there were also many unexpected challenges. There were aspects I thought would be easy that grew tiresome quickly, and obstacles I had never even considered. During the first few months, I wondered if we'd made a huge mistake.

From the first neighbors to drop by unannounced, to the worrisome realization that anything forgotten during my weekly grocery store forays would be half an hour away or more until the next time, country life required some changes in perspective that I had never expected.

Moving back to the land often carries romantic notions of cheerfully harvesting food straight from the abundant earth and caring for all manner of strong, healthy livestock. While that is the beauty of rural life, so are the months spent turning soil, planting seeds, plucking weeds, and battling off all manner of pesky and harmful insects. Plenty of dramatic hardships can occur on a farm. There is never a shortage of good stories that bring a sigh of relief in retrospect. But perhaps the biggest surprise of rural living is how totally routine it can be. And if you keep livestock, every day features the same monotony of morning feeding and putting them out to pasture, checking water buckets, breaking ice in winter, changing wet and dirty bedding, and lugging bales of hay. Animals do not like their routine altered, and they'll be feisty and unruly if you're a few minutes late.

This routine does not take a day off. One of the great lessons I learned as a child growing up with horses and chickens was that no matter how you felt, the animals still needed to be fed. Creatures that you keep as pets or livestock cannot take care of themselves; because of you, they are trapped in enclosures unable to feed themselves or keep their areas sanitary. So they depend on you, and they expect you at a certain time of day. With any kind of livestock, you do not get days off, you do not get sick days or vacation time unless you can find someone you trust to take care of them. This unglamorous routine goes on 365 days a year.

Even for the homesteader without livestock, the realty of rural life is a grueling workout. No one I know who has succeeded at homesteading spends any time not planning ahead. You may think you're out in a rural area where you can take long walks enjoying the sights and exploring the land. But if you want to survive, everything revolves around what you are doing to promote that survival: growing food, raising food, building shelter, or keeping warm. In order to thrive and provide you with food, a garden needs nourishment. In order to stay warm and survive the winter (and not risk burning your shelter down), wood needs to be constantly collected, cut, and stacked. A property needs continuous work, even with a fairly new construction, and there is always something to fix and improve on the land.

There's a comfort and a familiarity to performing the same tasks every day, and then the delight in the occasional out-of-the-ordinary experience: when the first doe delivers newborn kids, or a batch of fluffy goslings

Credit: Kate St. Cyr

Goats and other livestock need daily attention.

arrive. We humans are, after all, animals too, and it makes sense that we would want that kind of daily pattern. However, after two years on the homestead following the same drill every morning and every evening, monotony began to take its toll on me. I was feeling restless and missing the excitement that I once had from my daily tasks, and realized I was in need of a little bit of rest and recuperation.

I realized a break is necessary once in a while. And that is just one reason to build a support network around your homestead. Friends and neighbors, and folks living on other farms, can be very valuable if you need to step away from the farm in an emergency or for a break. These people are especially helpful in sharing advice and can commiserate with you over setbacks. Having at least one or two people that you can trust to take care of your farm when you're away, and turn to for advice if something goes wrong, is very important to surviving homestead life.

And there will be setbacks. There will be all of the problems that you've thought might happen, and more that are unexpected.

Everyone homesteads a little bit differently. Your manner of homesteading depends on your reasons for making this lifestyle choice. Some choose to live truly off of the grid, with no electricity or running water. Others, while living in a traditional home with heat and running water, focus on raising as much of their own food as possible and making everything that they consume. Plenty of people live somewhere in between, which is where we found ourselves upon moving to the country.

When we bought our farm, it had no electricity, no running water, and very little else. After weighing our options and priorities, we put connected electricity and water to the barn. We were fortunate to discover an old pipe linking the barn to a spring up the mountain, and simply reconnected it and added a hot water heater to provide an outdoor shower and functioning sink. In the house, we closed off two rooms for living, where we set up a bed and woodstove and kept a bucket for excrement, emptying and burying the contents twice a week. Our lifestyle was rough and unbelievable to many of my friends, yet it was the lap of luxury compared to many homesteaders I knew.

Even with our limited modern amenities, we still encountered unforeseen challenges. The first year, we put off winterizing the outdoor shower until it was too late and found ourselves bathing in the kitchen with a small tub of water and a wet towel for the next four months. Our first full summer, we relied on the natural spring for all of the farm's water needs. That dried up in August and did not give us another drop until November. No matter what you plan for, there are always going to be unexpected setbacks.

The Maine winters provided us with some particular challenges. We knew what to expect in terms of temperatures and snowfall, but planning to survive those elements is a lot trickier in a remote area. It is certainly true that the entirety of summer in a highly seasonal area is spent simply preparing for the upcoming season. Gardens are planted and harvested not for immediate vine-to-table consumption, necessarily, but for storage to use in the cold winter months when fresh vegetables are hard to come by. Heating with wood will have to start by cutting and splitting at least six months earlier in order to have it dry in time for use. Even if you're purchasing dried firewood, you'll be spending a good chunk of time in the summer stacking it. While the old adage says firewood warms you twice—once when you cut it and once when you burn it—I would propose it's actually thrice, since stacking is such a labor-intensive process. Depending on your location, winter may or may not be a major annual experience. If it is, events as basic as snowfalls can develop into logistical nightmares if you are not prepared.

There will also be drama on the farm. Sometimes things go wrong that can break your heart. If you farm with animals, you will have to address their mortality at some point. Many homesteaders raise animals for meat, and the reality of butchering your own livestock can be a harsh one. Even if you keep animals only for milk or eggs, accidents and illnesses can occur, and despite all your best efforts, at some point you'll lose an animal that was a great friend. Farming of any kind is never for the faint of heart, and since the rest of your livestock will be clamoring for attention, there is very little time to mourn on a farm.

The demands of homesteading can be divided into the physical and the emotional. Physically, you require a basic level of strength in order to be successful. While not every homesteader must be young and fit, you will be hauling bundles of firewood, heaving bales of hay, or bending over in the garden for days at a time. If you plan to keep livestock, you need to be able to physically handle your animals.

The bonds we form with our livestock can be deep.

Homesteading can be a great daily fitness routine. It calls for reserves of endurance, and same workout is usually without much variety for months at a time, changing only with seasonal requirements. That workout is demanding, and some of the most mundane daily tasks can be physically exhausting. For instance, when checking the garden for pests, I'm not carrying anything, but the act of bending over to check each plant puts a burn in your back, after working the entire quarter of an acre—and if you really want to be self-sufficient, a garden at least that size is recommended.

Firewood and water carrying are great muscle-builders for a farmer. Even on the grid with running water, you'll have to carry buckets for animals or gardens at some point. And there is no getting around firewood and the labors involved in its creation and use. If all goes well, the worst your body will suffer will be some blisters and sore muscles. You also need an emergency response plan should you be badly injured—which is very easy to do, especially if you keep large livestock—including someone to call on for help while you are laid up. Once again, neighbors or nearby farmers can be a tremendous help if you have built connections with them. If you are able to keep working with an injury, well, that is a hardship in itself.

The physical aspect of farming is especially important to consider if you are planning on this being your long-term lifestyle choice. Back in the day, farmers had large families to help keep up with the demands of their farm. Today it is less common to have so many children, and plenty of couples choose not to have children at all. Besides, having children on a homestead brings its own challenges. If you envision relaxing on your homestead in your golden years, it is possible. But you have to scale your life to conditions and activities that can be carried out when you are not as young and robust as before.

The emotional hardships on the homestead can be the most exhausting. One tragedy can bring you down in an instant, but

Caring for livestock is both physically and emotionally demanding.

more often a series of small hardships might work your system over until you feel like throwing in the towel at the end of a long season.

Even without mishaps, the mental wear of early mornings and late nights without a break can be taxing. Projects to prepare for the next season require time, which is taken away from any leisure or just added on as extra hours that you'll be awake and working. You cannot delay preparations on a homestead. Winter will come, and you need to be ready. If you keep animals or have a family, you owe it to them to be ready for every season.

The most effective way to maintain mental health and physical endurance on the homestead is to schedule your leisure time. In their book, *Living the Good Life*, Helen and Scott Nearing discuss the importance of evening meals with music and friends. Other homesteaders I know make it a point to get off the farm at least once a year. You can escape the monotonous daily tasks for a little bit, or even include leisure in your routine, but you will have to plan for it proactively.

Having to be mentally on, constantly thinking of solutions, can be as stressful as hard physical labor. Even on the smoothest running farm that you've been operating for years, there are fresh ideas to implement and new crops to try. On a new homestead, it is all about figuring out what works and what doesn't and making changes to ensure that you and your family are comfortable, safe, and achieving your goals. So while the homestead might let you rest physically occasionally, you'll find yourself unable to rest totally as you do the math for next year's breeding plans or crop rotations in your head over and over again, or try to figure out a better way to transport water or turn the soil without a rototiller.

For most homesteaders, the other emotionally exhausting aspect of the lifestyle is finances. While a few back-to-the-landers do back up their plans with money, the lifestyle generally attracts people working on a budget. And even with a financial cushion, you'll be amazed how quickly these lifestyle choices eat into any reserves.

As in any situation, it is important to keep an open conversation going about finances. For those trying to escape the "real world" on a homestead in the woods, you'll find it is impossible to begin without making initial smart investments. You won't be spending money on frivolities anymore: you'll be placing your dollars where you hope to recoup them on the farm or investing in tools or materials that will outlast you on the land. Maintain a budget whenever possible and research the financial realities of any animal or new equipment before adding it to the farm; try to make gains, not losses. All this sounds very practical and common sense, but to maintain focus and stick to a budget can be mentally draining—especially if you thought that was what you were escaping.

Money choices should be made wisely, but expenditures are well worth it if they help your lifestyle to be enjoyable and successful. Eating out, vacations, new clothes, all fall away if you're focused on your homestead. Instead, when you've got a bit of extra cash, it is invested in the farm, whether it be new equipment or new animals.

The best step to take to avoid emotional burnout or undo physical stress is to be prepared. Preparation takes its own mental toll and is hard work, but it is a lot less stressful that fixing a situation that has already gotten out of hand. While there will be plenty of unforeseen issues, consider what you already know. Every year you know winter is right around the bend, and every spring you should be ready for planting and new animals. There is no reason to make your life on a homestead any more difficult than necessary when you are fully aware of the cycle of the seasons.

My observation has been that many different kinds of people choose to homestead. Interestingly, while generally following the same lifestyle, they often couldn't be more different in their outlooks.

A "homesteader" is someone wanting to live in closer harmony with the land and be involved in the production of their own food. Homesteaders do not have to detach themselves completely from the modern world. They are excited about self-sufficiency and having control over their life systems and energy resources, but they are not restricted to an existence entirely separate from modern civilization. In other words, there are no "rules" for homesteading, nor an archetype homesteader.

I do not believe that a homesteader needs to make their life more difficult than necessary. Homesteading is about self-reliance, creating your own food and goods, and controlling your own life. It doesn't have to be detached unless you want it to be, and it doesn't have to overlook conveniences.

For example, do yourself the favor of getting the best quality tools. If you are going to be heating and cooking with wood, buy the best woodstove that you can afford. If you want to keep an extensive garden, find a good rototiller that you can operate comfortably. Some make the choice to till by hand, which works fine if you are prepared for that amount of effort. Do your due diligence in research before bringing home any animals, and continue to learn about their needs while they live on your farm. Invest in a good farm truck, or another vehicle, that can transport bales of hay and animals easily. With tools and animals, take the best care possible of what you have. Have a plan for all weather, although those plans always change. Within your own personal parameters, depending on your reasons for homesteading and how connected you wish to stay, get yourself the best gear whenever you can.

Homesteading is hard work. There is no break from the efforts. You deal with physical hardships, emotional heartbreaks, and you'll face the unexpected at every turn. Your ultimate goal, be it total self-sufficiency or simply some independence from the grocery store, will be the entire focus of your life.

Further Reading

- *The Contrary Farmer* by Gene Lodgson
- *Living the Good Life* by Helen and Scott Nearing

Questions Before You Leap

- How much land does the property have?
- What is the condition of the house and/or outbuildings?
- Does the property include any open fields or ponds?
- What is the condition of the land generally?
- How does that land's condition translate into your homesteading plans: for example, can you grow the crops you're planning on or pasture the animals you hope to raise?
- What level of off-grid living are you hoping to achieve, both initially and long term?
- Do you have a plan for obtaining a mortgage?
- Will you be maintaining a job off the homestead, and if so, what will your commute look like?
- Do you know and trust folks who can take care of the homestead if you need a vacation? If not, are you prepared to be on the homestead every day?

CHAPTER 2

Skills and Resources for Rural Living

Do you have what it takes to homestead? For some, the rural living skills required may be what is stopping them from taking the plunge. For others, the real-world survival skills that are fundamental to a rural lifestyle are part of what initially attracted them to homesteading.

Since "homesteading" can be what you make of it, you don't need to go completely off-grid to live a more rural and self-sufficient life. You may choose to give up all modern conveniences, or you may give up none but focus on producing what you use yourself. However, some basic skills will become necessary if you have chosen to move away from urban life.

Resources

Electricity

For many homesteaders, electricity is the first modern convenience to go. Living off the electric grid has its benefits, but before making that jump, it is good to think about exactly how much of your home runs on electricity. Simply put, it's not just lights.

If you have running water, electricity probably pumps it into your home and heats the hot tap and shower water. Electric heaters can provide a great way to keep a small space safely warmed in winter. Most modern entertainment systems run on electricity, not to mention ovens, stovetops, and fans. That water pump means that your toilet requires electricity, as well as your faucets, refrigerator and freezer, any computers or other devices, and an internet connection. You can cook and heat on a wood-burning or gas-powered stove, and you can do without computers or cell phones. There are certainly plenty of alternatives to electric heaters, and an old-fashioned icebox can help you enjoy fresh produce. But the lifestyle can be challenging enough without taking power out of the game, and there is no shame in homesteading on the electric grid.

You can also consider solar or wind power, or other alternatives to traditional electricity, instead of simply cutting that cord. Their installation can be expensive, but they usually pay off in spades down the road. Many states offer tax deductions and incentives for solar or wind-powered homes, and solar systems are getting cheaper every year.

If you want to cut the cord entirely, use candles and oil lamps to light your nights, the traditional way of living. For the fundamental off-grid enthusiast, there are always alternatives to the modern electricity-sucking option, and always more that you can do without.

Water

Water is the first requisite for survival. Everyone needs water. And if you plan to keep a garden or animals, you will need a lot of it. While you're looking at properties, finding a place with a natural spring or a stream running through is ideal—even if the land has a house with running water. It is always good to have an independent source of water, especially if you know your needs.

An on-the-grid home will have running water, which offers a great benefit. You may choose to add running water to an otherwise off-grid home for convenience, or carry in water for the majority of your home's uses. Carried-in water doesn't have to come directly from a natural waterway; off-grid homes can have wells outside the house that are used as freshwater for cooking and drinking. Either way, you'll be doing plenty of transportation. Water is tough to carry any distance, not only because it's heavy,

Our property has a seasonal pond, enjoyed by the ducks and geese in spring but dry by August.

but because it sloshes and leaks easily. Containers with lids are preferable. After a season of carrying five-gallon buckets to keep a garden watered, you will likely yearn for a method that uses less muscle. A tractor or the bed of a pickup works well, as does a hand- or horse-pulled cart for the human-powered homestead.

A natural source such as a stream or a lake may freeze solid in the winter months. So before you decide to rely on that for your water, have a plan for if it freezes. A natural spring should not freeze, nor should a drilled well. However, the pipes running water from a well to your home can freeze, as can the water pump—so make sure pipes are buried deep and pumps and indoor plumbing run through rooms that stay above freezing.

Not only is there the worry of having your home water source freezing in winter, the animals need fresh and unfrozen water year-round as well. Their water needs to be constantly de-iced throughout winter days and nights. To keep your creatures from catching a chill, it's good to serve warm water on especially cold days, and to refresh that warm water every three or four hours. Warm water will not only warm their bones, it will also take longer to refreeze.

Water for bathing, in order to be tolerable, needs to be warmed. For some folks living rural, this is as simple as turning on the hot water tap. For those living more off the grid, you'll find yourself warming water over a fire or on a stove.

Heat

It is my opinion that traditional oil heat should be the easiest modern convenience to let go of. Simply put, it is not difficult to keep a dwelling comfortable and warm without depending on supplies of propane, heating oil, or natural gas. Even with these heating systems, you can keep your bill low in winter by using a woodstove and leaving the thermometer turned down low.

Credit: Kate St. Cyr

A Vermont Castings woodstove keeps our small living space toasty through long winters.

Two options for off-grid warmth are a fireplace or a woodstove. A fireplace is a grand centerpiece, but it will only truly heat the room that it is in, it burns through wood quickly, and it won't stay warm through the night without stoking. Generally speaking, a fireplace also will not put out as much heat as a woodstove.

A large woodstove can keep a 2,000-square-foot space warm and also provide you with a cooking surface. Large cookstoves like this, generally made for the kitchen, can be very pricey. Smaller stoves designed strictly for heating smaller spaces can be perfect to keep a couple of rooms warm during the winter months. Woodstoves kept well stoked and fed dry hard wood will keep coals hot throughout the night and start right back up again in the mornings with kindling or small sticks of wood. Clean out ashes and coals regularly to keep your stove efficient, and make sure the chimney is in good shape to avoid any risk of fire. Depending on where you are, you can often get several cords of firewood to heat you through the winter for far less than modern oil would cost. If you harvest wood from your own land, your only expenses will be time and sweat equity.

Credit: Kate St. Cyr

A good carrier bag makes bringing wood in an easy chore.

For an example of how effective a woodstove can be, our farmhouse uses a Vermont Castings Resolute model woodstove for two rooms totaling around 500 square feet. We purchased it for $200 from Craigslist and installed it in our 1890s Cape chimney ourselves. With one midnight stoking and attention throughout the day, the stove keeps the space well above freezing, even when it's well below zero outside. When it ventures above fifteen or twenty degrees outside, inside it is a cozy seventy degrees or more. This level of comfort requires about five cords of dried firewood. We purchase wood already cut, split, and dried in May or June when it is least expensive. The stove burns almost constantly from October through April, and on a few chilly nights in summer.

Alternatives to full heating systems besides the woodstove include propane, kerosene, and electric heaters, and pellet or coal stoves. Electric space heaters are effective in small rooms, such as a bathroom, to keep pipes from freezing. Placed carefully, they present very little risk of fire. Propane and kerosene heaters can pack a punch of immediate warmth but present risks of both fire and carbon monoxide poisoning. While they are good to have in case of an emergency, I would not recommend relying on either as a primary heat source.

Credit: Kate St. Cyr

Loading up the woodstove.

In general, the placement of a heating device requires great care. Anything generating heat should be a safe distance from potential fire hazards.

Living Skills

Individual homesteads will come with individual challenges. Woods living is very different from prairie farmhouse life; off-grid living in warm climates presents obstacles not known to the Northern homesteader. From my experiences living in a small home on a windy hilltop in rural Maine, here are some of the skills I believe to be helpful.

Chopping and Stacking of Firewood

Please note: *I delve into quite a bit about woodstoves and firewood here. However, I am using anecdotal experience, and I am not an expert. Much more could be said about wood-burning, and you should consult an expert to clean and inspect your stove and chimney before using it. For the aspiring woodsman, there are many books dedicated to heating with fire, chopping and drying wood, and the whole world of using wood for heat.*

It seems pretty simple, but definite skill is involved in chopping and stacking firewood. If you purchase wood, it will probably come cut, and you simply have to make sure it is the appropriate size for your woodstove, or plan to cut it down to size if you purchase longer lengths. If you're cutting your own firewood, the first thing to understand is the process involved in getting it stove-ready.

When purchasing firewood, if you do not plan to chop it to size, make sure you measure what your stove can hold. Firewood is not a universal size, some stoves take 14", some 18", some 24", among many other options. Choose the best size for your stove.

A tree can be cut and sized with an axe or a saw, or a chainsaw and a sawmill. Once you have the right length, allow this green wood to dry for six months to a year. When it is dry, stack it for winter use. Ideally, the wood you're going to burn next should be

Types of Wood by Warmth

Wood	Million BTUs per Cord	Speed of Burn
Oak	24.0	Slow, a dense wood
Maple	24.0	Slow
Birch	20.3	Fast
Ash	23.6	Fairly slow
Pine	17.1	Fast, messy wood

stacked in an area close to the woodstove. A small supply kept nearby inside your house will prevent you from having to run out to fetch fuel from the main pile on a sub-zero night.

Keep small pieces of firewood and a hatchet near your woodpile. Small pieces can be used for kindling, and the hatchet helps you trim them down to good fire-starting size. Trimming pieces is easiest on a large log dedicated to that purpose, a piece forgiving to your axe that you don't mind being crisscrossed with chopping marks.

Properly stacked firewood prevents you from finding yourself on some cold day pulling a log out of the pile and having the rest collapse. Every good woodsman has his preferred method for stacking, but stability and ease of disassembly is key. When stacking for seasoning and drying, make sure the top layer is put bark side up, to prevent moisture from coming in.

If possible, stack wood in a shed, mudroom, or small outbuilding where it will dry faster, stay dry, and make gathering less of a chore. While outside wood can be kept dry with a tarp and, properly stacked, can season well in the elements, many homesteaders have found it worth their while to build a shed dedicated for wood as an easier way to keep it dry.

Tending a Woodstove

Even if your homestead doesn't use wood as the primary source of heat, you might consider having a woodstove burning in winter for supplemental heat or cooking. Tending to the stove is one of the most critical daily duties on a homestead. Fire is dangerous. As with heavy equipment or a large animal, always respect the possibility of an accident around a stove. Be careful and deliberate in your actions. If you do this, you should be able to use a woodstove for a lifetime without any issues.

Understanding your model of stove is critical. When you are first shopping for woodstoves, take into account the amount of space it can heat and its burn time. While you might think that one that pumps out a lot of heat is always great, it can be very uncomfortable and possibly dangerous if it is too large for your home. Burn time indicates how long you can leave the stove between tending, an important consideration for long winter nights when you don't want to get up every hour to tend the fire.

Read the reviews for the type of stove you are contemplating. Have your chimney inspected and cleaned every year for safety. Never overlook chimney repairs. If they are necessary, they should move to the top of the list of homestead projects. The same goes for any woodstove repairs. Understand how its dampers and flues work, and never leave a fire raging hot with the flues open for any period of time.

To start a fire in a woodstove, first make sure the flue is open. Crumple up some newspaper in the bottom of the firebox. Put a few small sticks of kindling on top. They can be cut from larger pieces of wood, scraps of bark, or unused shingles. On top of the kindling, place a few pieces of split, dried, seasoned firewood. Split wood is preferred to round logs for starting a fire because round logs take longer to burn.

Light the newspaper in several places. You can leave the firebox door open a bit to ensure good air circulation to get the fire started. Then, once it is burning, close the door. Use a special woodstove thermostat mounted on the stovepipe to monitor the temperature of the fire, and close up the dampers and flues when it reaches 300 or 400 degrees. Once the air flow is closed, the fire will slow down and burn to coals. Maintain the fire by opening up the flue and adding firewood periodically.

Regularly clean the ash box, if your stove has one, or allow the fire to burn down to ashes and clean them out from the firebox. How often this should be done varies on the model of stove, but it is crucial to keep it on a regular schedule. Store ashes in a metal bucket far away from the fireplace, preferably with an airtight lid. The bucket, if it cannot be emptied immediately, must be kept away from anything flammable. Dispose of ashes outdoors in a safe spot once the bucket is cool to the touch. Douse ashes with water or snow when possible.

Mechanical Skills

A totally off-grid homestead uses human and animal power to clear and maintain the land. However, plenty of modern homesteaders invest in a good tractor, rototiller, and other heavy equipment to help do these jobs. And if you've got equipment, you're going to need to maintain it.

Some of the most common mechanical tools on the modern homestead include a tractor (or two), a rototiller, chainsaws, building tools such as drills and circular saws, and a good pickup truck. You do not need to break the bank investing in these tools of the trade; many good deals can be found online or in the local paper for lightly used machinery that will serve your farm well. But even if you buy new, you'll find that equipment that gets used is equipment that gets broken.

I highly recommend a good basic understanding of any farm machinery. If you work as a couple, or with another family, at least one person should be able to repair or replace a part in a pinch. If

A tractor is a multi-purpose homesteading tool.

you don't have that knowledge (and the library is a great place to gain it), make it a point to find a mechanic that you can trust. Ask around, read reviews, and get a good feeling for a person; do not just find the closest guy to work on your tools. Chances are you'll be relying on that person, and you want to be getting fair deals and good work for your money.

Most homesteaders start out with little or no equipment and gather it over time. Hand tools are a great option to start with; they're less expensive and less complicated to work on. If you can sharpen a handsaw and an axe, you can start to clear brush or harvest trees. While this may not be practical or in the plans for every rural dreamer, starting out with a human-powered homestead can help you better understand what tools you need versus what tools you want. You don't need a yard full of equipment that never gets used—after all, equipment left to languish is all the more likely to end up rusted and broken down.

Good hand tools need care too. Those being used regularly can be sharpened and maintained on the go, but those used less often should get at least a six-month oiling and check. Make sure handles are seated well, blades are ready to go, and you have tools you can pick up and get to work with at a moment's notice.

Animal Husbandry

Contrary to a city-dweller's visions of a homestead, not every farm needs animals. You may choose to live for years without adding any animals to your rural life. However, animals go a long way in reducing a farm's grocery bill (even if you are vegetarian), and for many the appeal of farmyard friends is part and parcel for "going country."

The first thing I advise anyone following a rural dream is to think about what animals will benefit your farm. There's no use in a menagerie that serves no purpose and ups the grain bill. When you are first starting a homestead, the budget is often tight; you should work with animals that are within both your budget and skill set.

We always have more eggs than our family can keep up with and are able to sell extras to pay for grain.

The gateway animal for most is chickens, which are reliable layers and can also be butchered on the farm easily. The same goes for ducks, and I have a particular weak spot for geese. They aren't the most useful layers as their eggs are only seasonal, but these birds provide good protection for ducks and chickens and are excellent watchdogs. With poultry, you'll have animals that require relatively little maintenance and reward you with many benefits. Leave plenty of food, water, and a secure fence, and you can be off the farm for a day without worrying about them. The right breeds are cold or heat hardy, and it does not take any skill to collect eggs.

Small livestock is often the next step. Goats, sheep, and pigs are all great first homestead animals. They have specific space and feed requirements; you'll have to be home every day to let them out, feed them, and return them to their secure night shelter. Butchering, if you are raising for meat, can be done by a local butcher. Milking, if you're raising for milk, is a relatively simple process once you get the hang of it, though it requires a significant investment in time. If you wish to sell your milk, meat, or eggs, take into account your town and state's requirements for sanitation and sale, but you may find that is a simple way to provide extra income.

Sending animals to slaughter at a butcher's can be an expensive proposition; it's much cheaper to butcher at home. Without addressing the emotional aspect of harvesting animals, there is still much to consider in terms of how to most effectively kill the animal and how to harvest its meat quickly, cleanly, and efficiently. Adding animals for meat may seem to be a great way to feed your family, but at the end of a long summer bonding with your pigs, cows, goats, or chickens, you may find it difficult to kill them yourself. Faced with several hundred dollars in butchering fees, you may also find it difficult to take them to a shop. You need to be fully prepared for either situation when you first purchase the animal, otherwise you will find yourself with a pet instead of food.

Cooking

If there is one aspect of homesteading life you absolutely should do yourself, it's cooking. Preparing meals is one of the easiest ways to provide for your family, it brings immediate joy and satisfaction, and it can be done with minimal extra effort—even a "homesteader" in the city can utilize their kitchen.

The day-to-day processes of meal planning is a big part of homesteading. If you butcher meat or keep a garden, you're thinking far ahead in terms of meal prep. You plant vegetables for fresh summertime consumption and to preserve and store throughout the winter. Good food storage can keep you away from the grocery in the depths of winter, including putting plenty away through home canning. This requires a few days of intense labor in the kitchen, but you are rewarded with rows of colorful jars lining the pantry shelves, just waiting to deliver months of enjoying your own homegrown food.

Fresh garden produce

Food can also be preserved by freezing in some form or another and then thawed out for a later meal. The trickier part of food storage is aging or smoking, but even this is something you can tackle in your own home if you are adventurous enough.

Some people get into homesteading because of a love for cooking. The desire to include more fresh homegrown food in their meals and an inexpensive way to get unusual ingredients is a driving influence to "live off of the land." Others find themselves in the role of "cook" as a natural outcome of the homesteading lifestyle. Growing food in the garden, followed by the need to preserve and use it, sends the homesteader into the kitchen. After a harvest, this is sure to be one of the most rewarding parts of choosing rural living.

Other Aspects of Rural Living

Every homestead is different, and everyone has their own favorite aspects of farm life. If you purchase an overgrown property, an ample part of your schedule will be devoted to clearing out brush, which will require hacksaws and hard work, or chainsaws and clearing machines. All tools—especially gas-powered machinery—require personal attention, regular maintenance, and skill when being used.

If your dwelling and outbuildings were described as "fixer-uppers" in the real estate ads, then carpentry skills will be required. Even a well put-together home needs some know-how and living skills to make it comfortable throughout winter. In colder climates, winter preparedness is at least half of homestead life, and knowing how to meet the challenges of winter weather will make your rural life all the more sustainable and enjoyable. The best tip for surviving a frigid winter? Make sure any holes or cracks are covered and sealed, including closing curtains and making windows airtight with a layer of plastic. Small space or large, if you have a woodstove or other source of heat, and little or no way for the cold air to get in, you will be able to keep it warm throughout the coldest days.

Many homesteaders make their own soaps, skin care items, and herbal remedies or sew or knit their own clothing, and in other ways reduce the need for shopping and consumerism. Some remain completely on the grid, while making and crafting nearly everything their family uses. Others detach themselves completely from mainstream society or use an occasional grocery store trip to stock up on the needs that can't be provided by the farm. Regardless of your choice, you'll be learning new skills and utilizing talents you didn't know you had every day. There are no days off from homesteading, which is both its joy and its hardship.

Finding and Using Other Resources

For many, homesteading is all about self-reliance. But it is important to recognize where your skills end and to reach for help when necessary. When moving to a new area, take the time to get acquainted with your neighbors, visit the local businesses, and befriend the folks who you might be working with at the hardware or feedstores. The country post office is a good place to start. Learn who has what skills in town and what or whom to beware of.

Outside of broken machinery or larger-scale projects, locals can help you understand your new area. People who grew up there and whose families go back generations on surrounding properties will understand how your land works, how it has been used in the past, and may be able to offer great insight before you start planting a garden in an area that floods every season.

You can also use local knowledge to find good deals on equipment and livestock, but be careful and get to know folks before assuming that they are offering you a good deal. If you're moving in from an urban area, some will assume you are both wealthy and foolish and will be tempted to take advantage of you. It is good to have a basic understanding of how much things are worth so that you are not taken for a ride, and even better to build relationships first before asking anyone for any favors.

The internet can also be a great resource for the rural homesteader, connecting you with nearby towns; offering tips on how to fix, build, and cook; providing sales for animals and equipment; and even creating networks with faraway folks in similar situations. While I hesitate to say that anyone using social media is truly living off the grid, the networks online can be a truly invaluable resource. Not only can you troubleshoot problems with people all over the world, but you also may discover other farmers right in your backyard that you would not have otherwise.

Homesteading requires many skills, and just when you think you have what it takes, you'll be thrown a curve ball that requires a whole new ability. It is impossible to know what your new lifestyle will require of you, specifically, but gathering and reading books and continuing research is always a good idea. A homesteader is, if nothing else, a lifelong learner.

Further Reading

- *The Encyclopedia of Country Living* by Carla Emery
- *Norwegian Wood* by Lars Mytting
- *Canning and Preserving for Beginners* from Rockridge Press
- *The Modern Homesteaders' Guide to Keeping Geese* by Kirsten Lie-Nielsen
- *Raising Goats Naturally, 2nd edition* by Deborah Niemann
- *Storey's Guide to Dairy Goats* by Jerome D. Belanger and Sara Thomson Bredesen
- *Fresh Eggs Daily* by Lisa Steele

Questions Before You Leap

- What is your plan for keeping food fresh or preserved?
- Is your property suitable for solar or wind power?
- How will you bathe and get fresh drinking water?
- How will you heat your home?
- How will you keep your animals warm in winter?
- Are you familiar with and/or comfortable working a woodstove?
- Do you have a plan for firewood harvesting if that is a heat source for you?
- Do you plan to have any heavy machinery on your farm/ homestead?
- If yes, do you feel capable of doing minor repairs and maintenance yourself?
- Have you ever kept animals before?
- Are you comfortable with the idea of slaughtering your own animals for meat?
- How will your kitchen be set up (especially in a temporary living situation)?
- What is your favorite product or food to make for yourself?

CHAPTER 3

Earning an Income

The keystone of homesteading is self-reliance. Nonetheless, it can be an expensive lifestyle choice, and one that you will have to support with significant income. While at some future point a homestead may be entirely self-sustaining, usually at the beginning it consists of putting a lot of dollars toward something you hope will pan out into profits or breaking even. Every step of the way is an investment, and livestock, for example, can be hard on the checkbook.

Before moving to the country, think about how you are going to provide for yourself on the farm. Most homestead dreamers cite lack of money as the leading reason they are not ready to make the change. But it is still very possible to homestead with a good plan and a realistic budget.

Since the greatest cost in rural living is often the initial property, finding a place that is within your budget puts you ahead of the curve. We will talk more about finding the right property in Chapter Four. Keep in mind that, as in any aspect of life, homesteading is a balance of bills and income. You may find it less expensive than you first thought.

The cost of rural living is often lower than city living. Property can cost significantly less in the right rural location. The further away from civilization you are willing to move, the more land you'll be able to find for less investment. On the flip side, affordable faraway land means that any needs in the city will cost more in gasoline and you'll be very dependent on your vehicle. Equipment and vehicle repair costs add up quickly, and you do not want to find yourself stuck on a remote country road.

The cost of food, gasoline, and other living expenses varies a great deal from state to state, but it can be less expensive to live rurally. What can change when you go rural no matter your location is how many of those expenses you actually consume. You will most likely spend more on gasoline, yes, but your budget for eating out and delivery can drop dramatically. Most rural locations don't get any delivery food services, and the nearest restaurants may be questionable and are usually a decent drive. This translates into savings by inconvenience, but if you focus on saving dollars, it can dramatically reduce your food budget. Dining out can be one of the most nonessential expenses a family has, and by focusing on cooking at home, you'll start to save. Groceries themselves can be more expensive in the country, but with smart budgeting, buying in bulk, and growing what you can, you will still reduce your food budget.

As in any aspect of life, smart budgeting goes a long way to making country living affordable. Personally, without the pressures of city life and seeing people every day, I found myself spending almost no money on new clothes, preferring to shop at the local thrift store. We make everything we can, which is much easier to do than I anticipated. Vehicle repairs and equipment breakdowns are a number one expense, as is everything that goes into animal care. For ourselves, some things cost less and others are eliminated. It brings us back to basics, but it isn't as difficult as it sounds.

The purchase prices of off-the-grid supplies and appropriate appliances needs to be considered. Animals cost money in initial purchase, feeding, and housing; seeds for your crops can be expensive, and in reality, it is a far-flung idea to completely escape the monetary exchanges of the world. It is possible to break free with an established homestead, once you have items to barter and gardens established, but you will have to spend money to start your new life. So before you make the leap, make a plan for making it work monetarily.

You are likely to encounter a few unexpected and rude surprises, so having a good plan regarding the economics of your new endeavor can help keep you positive and provide a smoother transition for you and your family.

On the Farm

Initially it may not be possible to make your farm profitable, but often small farms and homesteads end up able to support themselves through careful planning and crops or products that are unique and exciting to consumers.

Vegetables

Perhaps the most common way a small farm can earn some extra income is through vegetable or herb sales. A roadside farm stand can provide some income in the summer months, but those serious about making a profit may take on farmers markets or CSA (community-supported agriculture) arrangements. Farmers markets, which are a great way to get noticed in the local community, do have up-front costs for the booth and participation. CSA programs usually have legal requirements and contracts to be met with your customers. The risk of a CSA is if you cannot provide, your customers will need a refund.

Spring turnips

While farmers markets give you a space to sell your wares in your local community, usually on a weekly or monthly basis and sometimes only seasonally, CSA programs allow customers to purchase a seasonal or annual share in your crops. You commit to providing them with a weekly or biweekly delivery of vegetables, meat, herbs, or other farm products in exchange for their up-front investment. Farmers often prefer CSAs because they can put customers' dollars to work immediately, growing the crops they have committed to selling. CSAs are also popular because they build a stronger bond between customers and farmers.

If you're just selling extra produce that you will not be consuming yourself, a farm stand is a great choice. Some become local attractions and can provide a fair amount of income, but looking into other avenues such as the CSA model is a good idea if you're really trying to grow and farm on a larger scale.

While vegetable, herb, or flower gardening may sound like an easy way to make money, it requires some initial investment and ongoing work. The cost of setting up a garden will involve cultivating and fertilizing the land, including having soil tested for the crops you have planned, and purchasing seed or starting stock. In the future, you may collect your own seeds, but the main

Harvesting snap peas, an early spring favorite

costs will be in labor—weeding, watering, and paying close attention to keep the plants healthy and salable. Keep a sharp eye out for damage done by weather, insects, and disease and react immediately. While many are able to garden and maintain a job off the farm, during the growing season a successful garden requires your time and attention. Like animals, plants need food and water to survive and thrive.

Even with hard work and a healthy garden, it can still be difficult to make growing and selling vegetables profitable, especially as more and more small farms turn to farm stands or CSAs to make money. Do not limit yourself to a traditional vegetable garden; see if your land is capable of growing unique plants. Then research the market and the demand for what you can grow, and you may find yourself with a profitable niche. Specialty plants, especially if not normally grown by local farmers, can provide more income either through markets or wholesale sales. While your risk is greater focusing on a single crop, your return can also be larger—and you can divide your farm between a few unique crops to limit the risk of a certain disease or a bad summer killing off your profits. Some New England farms focus on the saffron crocus, French lavender, certain species of mushrooms, and even bamboo, and make a tidy profit. That is to say nothing of the many fruits that can thrive and bring high demand to a small farm.

Eggs

Perhaps the easiest option for a bit of added income on the farm, eggs are a good reason for keeping poultry. Not only do chickens lay eggs, any farm-raised birds will, even if you aren't raising poultry specifically for them. Many states do not require any special licensing or inspection in order to sell your eggs.

For the farmer looking to make a profit from the egg business, you may wish to look beyond chickens. Eggs are a great way to break even on your chicken's weekly feed bill, but unless you keep a large quantity of specialty layer breeds, it is hard to get enough eggs to make extra money.

Goose eggs offer our farm a unique, though highly seasonal, egg harvest.

Profits can sometimes be found in more unusual egg production. Goose, emu, and duck eggs are uncommon enough to fetch premium prices from professional chefs and hobbyist cooks, although these larger birds may not produce as many eggs. This unusual poultry can serve as an added attraction for visitors. Birds such as peacocks and emus are sure to draw a crowd if you offer visiting hours—and their plumage is almost as valuable as their eggs.

Meat

Harvesting meat can be a good way to add to your farm's profitability. For any meat-eating family, keeping a certain number of beasts for personal consumption is a great way to reduce your weekly grocery bill. Those looking for larger-scale meat production will face certain initial costs, but hobby farmers harvesting and selling meat are more unusual than those selling produce and eggs, so the results can be more profitable.

The initial costs for meat production include setting up appropriate safe living spaces for your animals. Healthy animals require a lot more infrastructure than garden beds, including shelter from the elements, fencing to keep them contained, and food and nutritional requirements. Animals raised on pasture and able to forage as they would in nature will taste better and fetch a higher price, as shown by grocery store prices for corn-fed versus grass-fed beef cuts. This requires that you provide them with appropriate pasture space. Purchasing feed is a quick way to run up costs while keeping animals, as are requirements for extra feed hay and bedding.

Easily overlooked, there are also costs for selling and harvesting meat in a sterile, safe manner. Farmers selling meat will require special licenses and inspections to ensure that it is safe for consumers. Butchers can do the slaughtering and packaging for you, but can be expensive. You will also have to consider storage and shipment of your meat, which should be frozen unless it is going to be consumed quickly.

There is also an emotional toll to raising animals for meat. When first contemplating them, you may believe that the slaughtering process will be easy, but most often, it ends up being difficult for even the most steel-hearted farmer. It is not something you get used to. Talking about the good of humanely raising meat, not supporting factory farms, and thanking the animals for their sacrifice does not make it any easier to slaughter your own food. Many animals develop bonds with you and express distinct, fun, and curious personalities you will miss on your farm. It is not a reason not to raise animals for meat, but it is something to be aware of before you leap into doing so.

Milk and Dairy Products

Large-scale dairy may not be practical for the homesteader, but small-scale production is a viable option to make your farm profitable. A family cow can provide more milk than you can readily consume, so it stands to reason that three or four cows would give you enough milk to sell at farmers markets or a farm stand. Even easier to turn into a profitable operation are goats, relatively hardy small animals that require little space and produce delicious sweet milk that is also popular for easy-to-make (as well as some more challenging) cheeses and dairy-based sweets. Lower in fat, goats milk is sometimes recommended by doctors for young children and the lactose intolerant.

Goats are a particularly popular way to start a small dairy. In fact, 82% of goat dairy farms have fewer than 500 goats (NASS

Credit: Kate St. Cyr

A small dairy goat herd can quickly grow into a large-scale operation.

2007 Census of Agriculture), and like the benefits to raising chickens or other poultry for eggs, you can use them for meat, to breed and sell kids, or for fiber, while also milking them.

Of course, larger animals such as goats and cattle have more requirements in terms of housing and pasture space. Goats are notoriously difficult to contain, so there is a major up-front investment in fencing and shelter. For anyone selling milk, add in the cost in setting up a milking room and making sure that everything around the production of the milk, cheese, or other products is sterile and sanitary. There will be licenses and inspections involved before you can sell your dairy products to the public.

Producing dairy requires a regular daily milking schedule that usually begins before the sun rises. It also means breeding your animals every year and handling births and newborns in the spring. You will not get milk without breeding your animals, and you will have to decide how you'll handle the ever-expanding nature of a herd. Some small farms that pride themselves in strong bloodlines can make a tidy profit selling their offspring every year, others just continue to grow or use males for meat production.

One of the fun aspects of dairy farming is the many dairy-based products in addition to just milk and cream. You can make caramels and fudges, soaps, hard and soft cheeses, and many more. It's a great way to diversify your offerings to customers, or simply to find a niche that suits you perfectly.

Fiber

Fiber can naturally work into the picture when raising sheep or certain breeds of goats. They have many requirements in terms of spacing and shelter, and the process of shearing is a fairly intense one. It is necessary for their health to shear sheep once or twice a year. However, you can easily raise these animals for dairy or meat, and also profit from their good-quality fleeces, which can be sold as throws or rugs. The crafty homesteader might enjoy making homespun wool into textiles and articles to market.

Don't Forget the Vet

Animals require proper veterinary care to remain healthy. This means annual shots and check-ups, as well as special supplements and injections during pregnancy and after delivery. Any veterinary visit is expensive, and rural veterinarians will charge more for traveling to your farm. Be sure to include this when researching the cost of keeping animals!

Classes, Airbnb, and Using Your Space

You do not necessarily have to raise animals or plants to make a profit on the homestead. For many, the lifestyle of farming and homesteading is a desirable one that is just out of reach. With the right piece of property, you may generate income without all of the labor-intensive aspects of animal husbandry or farm life.

Airbnb allows you to rent either a single bedroom or whole home on your property to travelers. This can generate extra income with little effort. The space may require some remodeling, plumbing or electricity renovations, but often that's work you would want to do anyway. Keep the area neat, tidy, and friendly, and plenty of folks will happily spend the weekend away from the city observing the rhythms of your daily farm life.

One of the trendiest ways to bring folks to your farm these days are activities such as goat yoga. You can host a class on something farm specific, a skill you've learned through homesteading, such as making cheese, or you can focus on something totally different. Yoga, mediation, and other wellness activities can be done out in the quiet of a field, surrounded by your animals.

It might seem like daily life to you, but to many it's something currently unattainable, and guests will revel at the chance to be around goats, chickens, geese, and whatever else your farm has to offer.

A picturesque farm in the country is a great vacation getaway from the city.

Off-the-farm Income

Sometimes the easiest way to provide for a farm is by getting or maintaining a job off the homestead. In my case, and for many others, one person in the family maintains a full-time job, while another might do some part-time work and otherwise helps attend to farm projects on a daily basis. This is a one way to ensure that farm work is accomplished, while there is still an outside source of income.

Commuting versus Community

The downside to a job off the farm is the quantity of time spent away from your land. If you live in a truly rural setting, finding or keeping a job may require long hours away at the office as well as significant chunks of time spent commuting, transportation which could be costly and can make a such work impractical and unprofitable.

A job closer to home is always an option, but often it offers reasonable pay with good benefits but still requires a commute. The ideal farmer's job allows for working from home, which eliminates the time wasted traveling and allows you to be on hand in case of emergencies.

Do not assume that by moving to the country you eliminate all job options. There are often reputable small local businesses looking for qualified employees. Staying on at least part-time at a "normal" job does allow you to also maintain connections with your nearby town or office and a sense of community. While some of us prefer rural life to the social requirements of the city, plenty of others realize that the urban hustle and bustle of meeting new people is what they miss when living on the farm.

Whether you're able to maintain an existing job while going rural, or finding a new local job, a homestead cannot start out without some kind of income. Maybe a job is not part of your long-term plan to homestead, but at least for the first few years, that added income can keep you afloat and growing.

Bartering

A big part of country living can be bartering, trading, and helping your neighbors. It is a key part of homesteading on a budget, and a societal building block that should not be abused. If you have something to offer, be it manual labor, equipment, or farm produce, you can barter this for other equipment, labor, or produce from your neighbors. This simple age-old manner of commerce provides access to resources you might otherwise not be able to afford.

I believe that bartering and trading is especially useful for farming equipment that you would otherwise only need for a single project. You need it for a certain task but would not necessarily use it again. If a neighbor has that tool, maybe you can work something out.

It goes without saying that when borrowing equipment, always return it in as good condition—if not better—than when you borrowed it. Some neighbors may simply refuse to loan because of past experiences with broken down or never returned tools, and you should be selective in loaning out because of that risk. When bartering, always be fair and offer items that you know are of equal or greater value than what you're getting. And always offer something, even if someone has said they'll let you borrow an item without cost. It is better to offer a trade and maintain good relationships than to discover that that was the last item you'll be allowed to borrow for some time. Bartering means being a good neighbor.

Moving to the country can be expensive. Like any aspect of life, it is smooth going with a budget, realistic expectations, and plenty of financial planning. The first thing you will learn about homesteading on a budget is just how much you can do without and still be happy. If you can cut items out of your budget, you'll be surprised how much you do not miss them—especially when you have caring for animals and gardens and your farm dreams to keep you busy.

Further Reading

- *The Market Gardener* by Jean-Martin Fortier
- *Sustainable Market Farming* by Pam Dawling
- *The Farmer's Office* by Julia Shanks
- *You Can Farm* by Joel Salatin

Questions Before You Leap

- What do you have available to you to fund this endeavor?
- Do you intend to keep an off-farm job temporarily? Would you prefer to always maintain an off-farm job?
- What are ways that your farm can provide you with income? What might be unique small-batch items you can craft or harvest that would set your farm apart?
- Would your partner be able to maintain an off-farm job?
- If either of you are keeping up with an off-farm job, how do you intend to budget time so projects are still completed in a timely fashion?
- Does your property or buildings allow you "destination farm" opportunities such as setting up an Airbnb?
- What can you sell or offer in trade to other farmers for services?

CHAPTER 4

Finding Property

So you have made up your mind, talked it over with family and friends. Everyone agrees to the idea of homesteading. You've considered the financial angles and the various ways to make ends meet in the country. You're ready to accept the exciting challenge of making a livelihood from your own land, with your own hands. The next big step is finding the right property to call home.

The purchase of property is often the biggest financial hardship in taking the leap to the country. Making the right choice is critical to a successful life on the land.

The real estate search is often one of compromises. For example, if you'd like a truly large swath of land, you might have a hard time finding it right next to a major city (unless you are quite financially secure). Try to accept the idea that you can get your needs and wants met by making a few sacrifices until you find your dream homesteading property.

The Land

You may think you need a lot of land to start your rural life, but it can be surprising how far just a few acres go in a rural location. You can farm and provide for yourself on a reasonably small plot if it's used correctly and efficiently. Even then, the appeal of land to roam looms in your mind, and sometimes the expanse goes far beyond its practical uses. If you're adamant, then be prepared to look farther afield, far away from major cities and towns.

Our criteria for land was space for animals and privacy.

Clearly the condition of the land is important. You can purchase woods and turn them into gardens, as long as you are aware of your own abilities and limitations. You can plant or raise animals according to the land and its strengths, or you can search for property according to what animals or plants you'd like to raise. When compromising with overgrown fields you plan to clear or recently forested acres, be realistic about your capabilities. Clearing land will take months of back-breaking labor, often a project that will continue for years.

Homesteading is possible on almost any terrain, but your land will determine what you are able to grow and produce. The individual ideal vision of a farm is different for each homesteader. Maybe you do not feel capable of turning woods into fields, you might still focus on crops that thrive in shade or examine the option of hiring outside labor to clear your land. A rocky piece of property may be ideal for a herd of goats, but not so good for plowing and cultivating vegetable crops. Fields rich in one mineral but not another can be altered, or you can plant a crop that thrives in one condition and something else on another area that has different soil.

The most outstanding aspect to any plot of land is access to water. Having your own source of good uncontaminated, clean water is especially useful when your livelihood depends on growing crops or animals. Even if your prospective house is connected to town water that is sufficient for household needs, the availability of a small pond or stream on your land is even better, since you will want a watering hole for your animals. Check out the source thoroughly and have the water tested for possible contaminates before using it.

While land comes in all shapes and sizes—and remember that a homestead can thrive in many conditions—it is a good idea to set goals for what you want in order to narrow down your

We turned overgrown fields back to useable pasture by clearing and burning for six months before introducing our animals.

searches and not feel overwhelmed. The important thing is not to dismiss property outside your criteria, yet continue to search for something within those boundaries.

Our Property

We were looking for acreage and privacy primarily. The condition of the house was not significant to us, but with livestock ready to move in, we looked for a good barn.

The property we found has 93 acres of former pasture. Having been left untended for at least twenty years, the pastures required extensive clearing, but once the brush was gone, returned to hay quickly.

We have some timber on a mountainside above us and a spring-fed well but no indoor plumbing. The house had no electricity, but the chimney was in excellent shape, and we were able to install a woodstove right away.

The barn, a huge selling point for us, was a structurally sound 1890s building with more than enough room for all of our animals. It was a mess, full of leftovers from the previous owners, but once cleared out, it was easy to shape into what we needed.

The House

The first question to ask about the house is do you want one already built? Maybe you want to build yourself. You do not, however, want to end up in a situation where you thought you could complete a home by winter and you're still living in a tent in January, especially in colder climates. If you plan to build a home, either you need secure shelter while that process is going on, or you need a concrete plan to complete the project quickly, before temperatures drop and the snow flies. If you plan to stay in a temporary shelter for a prolonged time, you will need to have a planned heat source and reliable water supply. And remember, many times temporary becomes permanent: try to stick to your long-term goals.

There are, in essence, three types of living situations for the new homesteader:

- No solid building, using a temporary shelter (such as a yurt, RV, or tent)
- A house in need of work, with ongoing projects perhaps lacking some modern amenities
- A completed traditional house

Our house offers quaint New England charm at the front door—but inside, it requires a full renovation.

Although a parcel of land with no house can be maintained while living elsewhere, this requires a monetary investment—financing and keeping up two properties instead of one. Plus, you will be commuting to your land, and that takes time and money.

If you find the perfect piece of property with no house and you want to move immediately, there are still options. You can put up a camper, a yurt, or other temporary home before you think about long-term housing. You will still have to consider the investment of water and heat resources.

A temporary house plan can work. Be aware that this might seem like an easier solution than it is in reality. Your supply of water and electricity, if you plan to use it, as well as heating and cooking still need to be addressed. Laying in water or electricity is a major up-front cost, even if you have no plans to build a house around this new infrastructure. Temporary structures can be

Credit: Kate St. Cyr

Our temporary bedroom setup isn't pretty, but it's functional. The fan in the doorway pushes heat from one room to the other.

unstable and more difficult to keep warm than a well-built solid house; they can be difficult or impossible to insulate depending on their material. Setting up a living situation like this, while it might save money in the short term compared to building a home, is often a much more expensive proposal than expected. And if your goal is to build a permanent home, your finances are instead going into the temporary one.

You can also find a property with a home that needs a lot of love. If you are homesteading, the house isn't going to be your only priority, and it won't be where you spend most of your time (with the possible exception of the kitchen). You can make do with a home that needs attention, and you probably won't be bothered by the idea of keeping up with the neighbors as you might in a suburban community. Home repairs that are not chronic can be addressed over time, as budget and schedules allow, and are often delayed. Accept that the house will be in rough shape for a while, but with determination and a good plan, it is possible to make it work out successfully.

An existing structure may already have electricity or water, saving you a boatload of cash. If it doesn't, at least there is more to start with than virgin land. You will, as with a temporary structure, concern yourself with water and heat, and how you'll cook. Just because a house has walls doesn't mean it is insulated or will be warm. Look into tasks like installing insulation, sealing off the windows against drafts, and placing hay bales around the foundation of the home to keep it draft-free in winter. If you are using any existing chimneys or other heat source, have them properly inspected before using them.

The fixer-upper is often the preferred housing for a homesteader. Property with an "undesirable" might seem like it's merely land without a shelter at first; however, it allows flexibility to either build or remodel your dream home. Pick and choose your priorities in terms of both remodeling and installing modern amenities and appliances, preparing a secure and warm place to live through your homestead's development.

Outbuildings and Barns

Equally if not more important than a house to any farmer is a barn. A structurally sound barn can raise the price of a property, but if you already have animals and need a good barn as soon as you can move, then the cost is worth it. A barn can serve many other purposes: transformed into a temporary living space, used to store possessions that you can't currently house, or used to shelter delicate equipment from the elements. In short, if you can afford to search for a property with a barn, it is well worth the added investment.

Barns come in almost as many different shapes and sizes as houses. What you want to look for beyond the number of stalls or the size of the haylofts is the condition of its frame. A barn with a rotted skeleton is of little use, if you end up constantly repairing it and spending as much as you would building a new barn. A barn with shingles falling off and some broken windows that has a good frame can be a manageable investment.

Credit: Kate St. Cyr

The barn with its strong bones was a big factor in choosing this property.

Any outbuildings, barns, or chicken coops are going to be helpful if you plan to keep animals. Of course, you can build new structures for them, but an established barn can save time and money if it is the

The most major area of work in our barn was the floors, many of which had to be replaced.

right space. In harsher climates, a full barn goes a long way in protecting your animals from the weather, and even a small barn can offer plenty of storage space.

Unfortunately, many barns in rural areas are structurally unsound, some simply falling back into the earth. You may salvage materials from these structures, but they prove to be inadequate for keeping out wet weather and high winds. Sometimes such frail structures are not even safe for housing animals. If you find an affordable property with a good stable barn, it is not a place to pass up.

Location, Location, Location

Often the first conversation before going rural is where you want to be. While a spot near a city can be more expensive, plenty of would-be homesteaders prefer to be within driving distance to their former lifestyles.

Because homesteading provides such a close connection with the land, you'll want to think about the climate of your new location. It is often nice to stay within a region with familiar weather patterns, because you'll have a greater knowledge of how to combat their extremes. If you do want to move to an entirely different zone, make sure you do plenty of research and be prepared for the weather. This could be your biggest obstacle to a satisfying transition.

If you live in a coastal state, you probably know that coastal property is desirable and usually the most expensive, but go an hour or two inland and prices fall. Even in a land-locked state, the farther you circle away from established towns, the more land you will find for lower prices.

Obtaining a Mortgage

How are you going to pay for the property? Land is the greatest initial investment for a homesteader. You will have to either put a large sum of money down or work with a bank or the seller to finance a mortgage agreement.

The problem with traditional mortgages is that banks are sometimes reluctant to finance the types of properties homesteaders are considering. No running water or electricity? No mortgage.

If you are able to pay in full for a property, you will be ahead of most folks, and you'll be staying out of debt. It is not always feasible, but if you can afford this, I would recommend it. You may be putting all of your eggs in a proverbial basket this way, but it is often better to own the land outright and work on the structures and lifestyle as you can afford to than the other way around. You may not have a lot left in the bank, but you will not be in debt and can put all of your earnings toward improving the property.

You can sometimes work out creative financing with the person selling the land. These arrangements, which are irregular but not unheard of, depend on the flexibility and willingness of the owner, as well as your comfort level. In any situation like this, always get your legal paperwork in order before proceeding and ensure that each party trusts the other.

There's also the option of working out an agreement with a traditional bank. Just because a home mortgage may not be possible for you, it does not mean that a loan is unlikely. Many banks offer a building or construction loan for a property that requires work. Their only concession is that these agreements usually have a short time frame for paying it back. So be prepared for that large bill to come due within three or five years.

Farming Someone Else's Land

Where I live, an organization called the Maine Farmland Trust connects farm owners who can no longer manage their lands with people eager to work the land but unable to afford it. The Farmland Trust also puts farms in easements and generally encourages the preservation of productive land.

Not every state has a similar program, but many do, and where no formal organization operates, there are still farmers looking for help with their land. It may not be your dream to manage someone else's property, but it is a great way to start farming without the huge initial cost.

There are many levels of farm caretaking. You may be managing someone's farm but not directly involved in its long-term plans. Or maybe the owner is allowing the farmer/caretaker to run their own small business or homestead on their land because they want the property to be put to good use. A relationship like this might be just what you want: the experience of rural living without the long-term commitment and out-of-pocket investment.

This arrangement gives something back to the folks whose land is not being used, and when it is successful, everyone is better for it in one way or another.

Further Resources

- Agrarian Trust: An organization dedicated to helping the next generation of farmers gain access to farmland, agrariantrust.org.

Questions Before You Leap

- How much land do you want and/or need?
- Where do you want to be?
- Do you want woods, open space, or a combination of the two?
- Are you comfortable living in a yurt or temporary home, or building a house, or would you prefer the home to be ready to move into?
- How far off the beaten track are you willing to go?
- Do you have the skills and time to build or rebuild a house, barn, or other structures?

CHAPTER 5

Infrastructure and Equipment Basics: Sorting Needs from Wants

Homesteading is about simplification. Many wannabe homesteaders are afraid to make the leap because of budgetary restrictions, but it is surprising what you can do when living off the land. Some things you will want to splurge on, and plenty of others that you can problem solve without overspending.

The House: Appearances Aren't Everything

If you are living in a traditional house, with modern amenities or not, home improvements can largely be made secondary to land and farm improvements. As long as a home is functional and structurally sound, it can be made to work. If you are homesteading, you should not be worrying about impressing the neighbors with your home. If you have neighbors or family come to visit, take pride in the big picture of what you are doing: providing yourself and your family a living from your own land and ingenuity.

The property that we found, while it had an almost ready-to-use barn (once all of the previous farmer's refuse was removed), had a house that hadn't been lived in since the 1970s. The small square Cape Cod-style building had a rambling addition that was falling apart, and inside the more structurally sound original part, the horse-hair plaster was coming off the walls and the floors were buckled with age. Without the money to rebuild or build new, we decided to make do for the first few years of our homestead life. We repaired the plaster where we could, closed off the area that was in the worst state of disrepair, and spent a week scrubbing and painting so that the front two rooms looked almost new. There was no plumbing and limited electricity provided by an extension cord from our on-the-grid barn, but it was cozy and clean. It wasn't high living, but it was perfectly functional, and more of our time and money could go into the barn and land.

It is perhaps impossible for a person living unhappily with a flush toilet to imagine a person living happily without one.

Wendell Berry, poet, activist, and farmer

Plumbing + Running Water

One of the easiest amenities to do without on a homestead is a traditional flush toilet. It can be surprising how much you do not miss it. Putting a flush toilet into a home without modern plumbing is quite an undertaking: it requires not only water pipes, but a leach field, septic system, and other expensive steps before you can think about which porcelain appliance to buy.

One of these options can replace the conventional flush toilet:

- Composting toilets are available for a reasonable amount from many retailers.
- An outhouse can be built, which can be a simple undertaking or can be quite elaborate and comfortable.
- A very simple camp toilet can be set up using just a five-gallon bucket and pine shavings.

Waste, once the bucket is full or the outhouse needs emptying, can be buried, or it can be recycled as human manure. If you're using an outhouse, it can be raked away and spread across your fields.

Water for drinking is a more critical issue. If you want hot water available from a tap, it is even more complicated. While you can harvest water from a local stream or pond, be sure it is safe for drinking before you consume it. You may also purchase bottled water or draw water from a neighbor's property or other location, but that limits your self-reliance a good deal. If using one of those options, bring jugs and large containers to fill up weekly or daily and take back to your homestead.

Digging or drilling a well and running a line to a house or barn is a more conventional option. For most areas, it costs over $5,000. If you need a reliable source for animals, running water is a definite asset. Even if you only have pumped water available to one structure—barn or house—this fundamental need will make your life more comfortable, convenient, and healthful.

Like a toilet, showers are somewhat optional. A person can stay clean for months with a wet towel for washing and a basin for rising sensitive areas and hair washing. We have piping for an

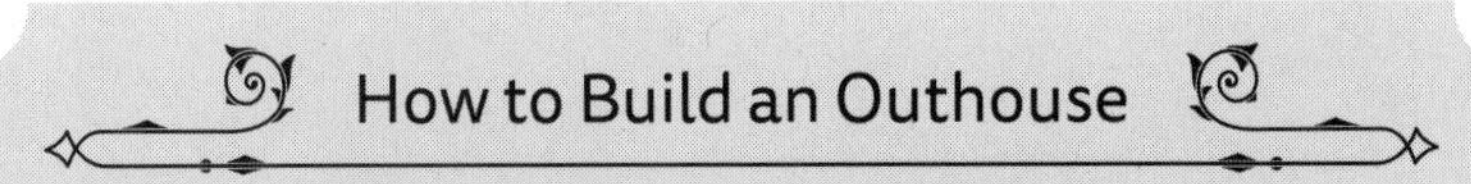

How to Build an Outhouse

- Before building, study location, distance from house. An outhouse should be on a downgrade from any gardens or animal spaces, as well as your home and water supply, and at least 100 feet away from any water supply.
- An outhouse pit should be around nine feet by nine feet and at least five feet deep.
- In addition to building walls and a roof (a roof being crucial for using the outhouse in bad weather!), consider planting a hedge of taller bushes around the area, or putting a bed of nice-smelling flowers nearby.
- An outhouse should be well-ventilated, but cover openings with screens to stop flies and other insects getting in.
- Wood shavings are best for covering odors and breaking down materials; ashes or lime also work.

Laying the water lines for our outdoor shower. We use an air compressor to blow the lines out in winter, and bathe when it is below freezing in our little kitchen.

outdoor shower for summer use. It has hot water, but we aren't able to keep the pipes, which run outside, thawed throughout winter. Without any piping, you can still enjoy bathing on sunny days using sun showers, which use solar energy to heat water. They are easily rigged and available for purchase from a camping supplier.

Heat

We've talked about woodstove heat previously in this book. If you have power, portable electric heaters can also provide good warmth for small insulated spaces, and they do not put off any dangerous fumes. An electric heater can keep a space about a hundred square feet above freezing. Electrical hookup is required, and keep in mind that they use a great deal of power. If you are on the grid, your bill will be quite high while using a space heater.

You can forgo the electrical costs with kerosene or propane heaters. These require good ventilation, or you can risk death from their toxic fumes. They can be handy for getting an area toasty in a fairly short amount of time, but they shouldn't be used exclusively or for long periods of time.

Putting in modern heating systems can provide some optimal creature comforts, but the up-front cost is fairly expensive; this is a good option for auxiliary heat. Even if your house has modern heat, you'll probably want to keep your thermostat set low in the winter to save funds. I do believe that a woodstove is the least expensive and easiest to maintain and run yourself option for a homesteader. That is not to say that installing a woodstove is cheap, nor is its upkeep simple or something you can overlook; it is just the best of the options available today.

Lights and Technology

Flipping the light switch and having your home suddenly illuminated might be the most presumptuous aspect of modern living.

If you don't have modern light fixtures, it's easy to get away without them in today's world. LED lights run on batteries that can be replaced or even charged with solar energy during the day and provide light at night. You can use older lighting implements like candles and kerosene lamps, but always be aware of the fire hazard. Properly cared for, these cast lovely gentle light, but you need to keep up with cleaning and maintaining them. Lamps will need kerosene fuel and regular wick trimming.

In today's world, we also find ourselves worrying about how we will charge our devices. If being online is important, figure out how to connect without building up data fees. Thankfully there are plenty of solutions for this as well, including solar charging stations.

Our kitchen is small but functional and cozy.

No Kitchen, No Problem

You also do not need a designer kitchen on the farm. Even if you are contemplating days of vigorous canning and preserving produce from your garden, or the daily maintenance of milk production with dairy animals, you can survive happily with a makeshift setup for your kitchen. You will still end up with delicious produce.

My personal experience is with a 100-square-foot stall in a corner of our barn. With a bit of insulation on the walls and some trim around the door, the room has been transformed into a cozy living space. We already had electricity in our barn. Fortunately, our property had a natural water source already piped to the barn from a spring, and we just had to install a deep laundry sink and hook it up. We added the refrigerator that had been abandoned in the house and were surprised when it whirred to life. It probably costs us more in electricity, but the fridge itself was completely free. We didn't need a great deal of fridge space for two or the fanciest new appliances as long as they kept our food fresh. With a couple of outlets added, we were able to include hot plates and eventually got a small countertop oven. A table and a few chairs, again found in the old house, completed the space. It doesn't look like a slick showroom farmhouse kitchen, but it is absolutely functional and we enjoy many home-cooked meals there.

We also have the option of adding a portable propane outdoor cooker to boil large quantities or for major canning projects. I can also fire up a grill whenever I want or need to. And for a delicious smoky flavor, there is always the fun of cooking directly over a campfire. If you have a woodstove, you can boil water, fry eggs, cook a quick stir fry, or bake a batch of cookies just as well as anyone with a modern gas or electric range.

The Barn

Choosing Animals

Raising livestock can be profitable, contributing to your plans of self-sufficiency, but after the significant up-front costs of purchasing them and providing shelter and pasture, you will have to add their feed and veterinary expenses. For most livestock, paying an initial premium is worthwhile for getting good bloodlines and raising healthy animals in the future.

The easiest way to avoid the expense of animals is to abstain from adding them to your farm. Yes, they're cute and fun and useful, but do you need livestock on your farm? If you can avoid adding animals, you'll save a lot of money immediately and have no veterinary bills to worry about or bedding or hay to purchase. And you will find yourself with a more flexible, free schedule to do whatever work is required to provide for your homestead and keep everything else on it functioning.

Yet, for some people, the animals are why they farm. Our own little homestead would probably exist in some capacity without

Geese are a large part of why we were ready to move to a farm in the country.

animals, but it was our growing flock of geese that prompted our move to the country. And I do not see our herd of goats diminishing anytime soon.

If you decide to include animals on your farm, try to be smart about what ones you choose. Poultry is the cheapest to buy and to raise, and will reward you with eggs that require minimal effort on your part. Collecting eggs is a pleasant chore, and they can be quickly consumed or sold. Goats can help clear land as well as provide milk if you breed them, while cows and horses require more space and time to care for. Start small and work your way toward more animals if that is your goal.

Animals = Secure over Picturesque

For the homestead with animals, remember that functionality is more important than appearance. You want safe, secure housing for your animals. Do not compromise when it comes to these priorities. Safe animals are happy animals, and healthier, happier stock create good-quality produce and offspring, resulting in better profits and a level of satisfaction and pride for you and your family.

You can skip the most picturesque white picket fencing or a grandiose barn, but you have to make sure that your animals have adequate housing. Any livestock needs a space secure from rain, snow, hot summer sun, and harsh winds. What is required, at its most basic, is four walls, a good roof, and warm bedding. This shelter can be a converted shed space, a coop put together with recycled materials, or even a repurposed area in a garage, as long as it has proper ventilation. Goats are sometimes housed in large doghouses, and plenty of chickens overwinter in hoop greenhouses.

What you want is a space that can keep your animals out of the elements. The flaw with a doghouse, for example, is that it will sit directly on the ground, and once the earth freezes, it will be almost impossible to keep warm. Goats don't need a really warm space, but they prefer temperatures above zero; in prolonged cold, lying against the frozen earth will affect their health.

It is usually easier than you think to build housing for your animals, and less expensive, as long as you keep a budget. Any existing or purchased storage shed can be made into a coop or tiny barn with a few pallets added for walls or nesting boxes. Look for salvage material on websites like Craigslist or use extra boards from your other homesteading projects. I would strongly recommend installing a floor under any animal housing for the reasons of frozen earth mentioned above (and for easier cleaning). While wooden floors will eventually rot and need to be replaced, they're often the best option. Cement or a smooth dirt floor can be cold and wet on long winter nights. Make sure windows and doors can close snuggly against the wind.

It is generally a good idea to separate species, especially those with a large size difference, such as between chickens and geese. Bullying may occur if the smaller animals don't have a private area for sleeping. Some animals may not be able to eat the same feed as others, or require certain minerals in their diets. Each has its own distinct needs and preferences for housing, so be sure to do your research before starting construction. Keep in mind that you can get creative if you wish, and you can almost always used recycled materials. The point is to ensure a stable, efficient shelter for your animals. You are responsible for their health and safety.

Likewise, do not skimp on fencing, both for your animals' protection and to keep them out of your garden and crops. Remember, both housing and fencing, while they required substantial costs, are features you are likely only going to have to do once. Fence posts can be made with trees from your land, if you have any that are sturdy enough; cedar is the most recommended. Post holes can be dug by hand or using a manual post-hole digger, and then your only store-bought item will be the fencing material. Better still, you can construct a post-and-rail fence entirely by hand with your harvested trees—just keep in mind that what you'll save in money you will probably be spending in time on that project.

However, post-and-rail will only contain certain animals and certainly not goats or high-flying chickens. Consider the escape talents of your animals and ask other farmers who keep the same types on their farms. For goats, I always recommend no-climb square-hole wire fencing; for chickens to be properly contained, a run with a roof of bird netting or chicken wire keeps them in and predators out.

Equipment Choices

After land and animals, your next major expense is often equipment. Even if you plan to do everything by hand with human power, you will need tools to complete tasks.

If you plan to have a garden, figure out how you want to turn the land. There are a few levels of tilling. Some folks, especially

Rototilling the garden in spring with a walk-behind rototiller.

with smaller garden beds but also on much larger areas, do hand cultivate the earth with a simple, sturdy pitchfork and a lot of time, patience, and muscle. Several gardening companies sell a broad fork, a kind of oversized pitchfork with knife-like teeth to cut up the soil. Still other farmers use a hand tiller, a tool with a large front wheel, a cutting blade behind the wheel, and handles so it can be pushed along. This is also an effective for weeding paths.

For those turning new earth, a tiller with a motor is often preferred. Walk-behind rototillers, which usually cost between $500 to $1,000, allow you to turn large swathes of land. They can still be exhausting to use, but not nearly as exhausting as using a fork, especially over a large area. Rototiller attachments for tractors make the job a song, but in addition to costing well over $1,000, they require that you own or have a tractor.

Purchasing some kind of powered rototiller is a good option if you're turning new land, and especially if you foresee adding new garden beds every year. However, if you just need to start one small area, you may want to look at hand options. Not only are rototillers expensive, but continually turning the soil to the same depth every year harms the tilth—the condition and tillability of the soil. Rototilling annually can harm microscopic life that lives in the soil and aerates the earth, providing the healthy soil you want for growing a garden.

Once a garden has been tilled, you can keep the soil in good condition by mulching heavily. Using straw or hay or waste from your animals' stalls, you can build up protection over the soil during winter and simply remove that layer in spring.

Our tractor has proved its many uses, including transporting shingles during the project of residing our barn.

A tractor is a useful piece of farmyard equipment. The price point on this classic farming tool can range between $2,000 to well over $25,000, and you do not necessarily get what you pay for. Before you start looking for a tractor, think about what you would want to use it for and what attachments you need or might want later. A bucket scoop is a helpful attachment, as it can be used for plowing snow and moving manure piles, as well as transporting any number of objects.

You do not necessarily need a tractor, however. Other tough and versatile equipment can accomplish many of the same tasks, such as utility vehicles with small bucket attachments and beds that can transport hay. A beater pickup truck can work to transport plenty of heavy equipment, but like any motorized equipment, it requires maintenance; you do not want it to break down just when you need to use it.

A simple pickup truck is one of the most valuable pieces of equipment on the farm. It can get you to farmers markets and other events, it can also haul hay, transport livestock, and take you off-road and across your fields. A newer pickup is an expense but can be worth it for the reliability and larger beds available. Even much older models can be reliable transportation; the fact remains that a pickup truck will need to be the most reliable tool on the homestead.

If you are cutting your own firewood or clearing land, you'll want some serious brush cutting equipment. A chainsaw is the most effective way to fell a tree or clear thick brush. A good chainsaw is a reasonable investment at between $200 and $700. It needs to be cared for and sharpened to perform well, and you must be comfortable and trained in its use before going alone into the woods. You should also spend the money for your safety and invest in good gloves, chaps, hearing protection, and safety glasses.

The alternative for the human-powered homestead are good axes: a large splitting axe, as well as some smaller ones for chopping kindling. A maul and wedges will help when cutting a large stump into sections for firewood. It is a good idea to have these hand tools around regardless of how you fell larger trees, especially if you're going to be sizing wood down for a stove. A large crosscut handsaw can also fell a tree, or cut branches into pieces, and it can be easier to use than an axe.

There's a myriad other tools you'll find yourself reaching for on a homestead. You'll need a drill and screws for endless tasks, as well as a good hammer and nails. A large wheelbarrow or wagon will see plenty of use, even in addition to the bucket scoop of a tractor. You will always want a good pocketknife in case of any emergency that requires cutting line quickly, or generally for opening up hay bales and the like. Wrenches, ropes, extension cords, ladders, shovels, a crowbar, and several good water hoses are used on a daily basis on our homestead.

Bartering

If you cannot afford new or expensive equipment or stock animals up front, consider the most ancient form of commerce. Bartering is one possibility for you, and a good option if you need certain tools just once or twice a year.

If your farm produces goods that can be offered as a bartering tool, try it. Your own labor or any equipment that your neighbors do not have are also useful options. You can barter not only for the use of a tractor for a weekend to get your garden tilled in the spring, but also for a new goat or a flock of chickens. As long as you are up-front and honest about what you can offer someone in exchange, a bartering system should work well for you, and trades like that can be something you rely on for certain work or equipment.

Used Equipment

The best way to keep up with the demands of the homestead is to not limit yourself to getting new products at the store every time.

It can take longer and require some more patience and research skills, but reliable used equipment can be found online. Beware of scams. Do plenty of research on what you are buying and always view the item before purchasing if it is possible. You widen your options the farther if you are willing to travel for goods, but often appropriate equipment can be found nearby. And nothing beats word-of-mouth. Next time you are at the feedstore check their bulletin board, or mention to the proprietor what you are looking for. You may come across a deal you would never have found otherwise.

If you're homesteading on a budget, do not limit purchases to the newest, or even the newer, models of equipment. In fact, when it comes to tractors, older models are often easier to maintain, which makes them a much better fit for the self-reliant homestead. Some items are worth buying new, especially when current

Credit: Kate St. Cyr

Most of our cooking needs sit on a single shelf, except for the pans that hang nearby.

models aren't that much more expensive than used ones. But for rototillers, tractors, trucks, and other big gear, pre-owned is the way to go.

Doing Without

The easiest way to save money is to do without. When we first moved to our country farm, I was concerned about how many items we were stuffing away into storage. We had come from a 1,700-square-foot, three-bedroom home, and we were filling a storage room to the gills with furniture, light fixtures, formal clothes, books and DVDs, kids' toys, and pots and pans. I was left with using a total of two cast iron skillets for the majority of my cooking, and the only furniture we squeezed into our living area was a bed and a bureau.

Since then, as we've grown used to the space and found extra storage places, we've figured out what works versus what gets in the way. But it didn't take long for me to realize I was not missing everything. Within the first week, I was comfortable cooking and living in our new downsized style, and now I can barely remember what is tucked away in that storage room. When we have the space to bring out some of our old furniture, I am guessing that a yard sale will be in order.

The fact is that we spend our time and money on the land, the animals, and the barn, and we rarely lack for creature comforts in the house. That doesn't mean that everything is always perfect, but the things that matter are where we've put our time and attention—and that care has proved worthwhile. Our hard work is reflected back to us in the obvious improvements that show up year after year.

Further Reading

- *The Organic No-Till Farming Revolution* by Andrew Mefferd
- *Essential Composting Toilets* by Gord Baird & Ann Baird
- *Essential Rainwater Harvesting* by Rob Avis & Michelle Avis
- *Horse-Powered Farming in the 21st Century* by Stephen Leslie
- *How to Restore Ford Tractors* by Tharran E. Gains

Questions Before You Leap

- What kind of equipment (tractors, rototillers, etc.) do you have to start with?
- What can you do without in terms of modern amenities (toilet, shower, etc.)?
- What does the property have to offer in terms of animal housing and fencing?
- What do you feel capable of installing in terms of animal housing, fencing, and garden space?

CHAPTER 6

Creating Community

The hardest aspect of "going country" for some people is leaving an established, supportive community behind. But country living can actually create some of the strongest bonds with friends and neighbors you will ever know. Away from cities and towns, people rely on each other, whether for help with day-to-day projects or in emergencies. While not every neighbor will be your closest friend, certainly a good neighbor is worth their weight in gold.

When you first move to the country, it can be difficult to form friendships. You may feel like an outsider who has left another life behind. For some, that is precisely the allure; for others, it is something to overcome. In many cases, you may be expecting a life of solitude, only to find an endless string of curious neighbors knocking at your door welcoming you to the community.

Neighbors

The truth is that you are likely to feel uprooted and lonely after moving to new territory. These are normal and natural feelings; just keep in mind that you are starting a new adventure and you are only on the first page. Having said that, country living can be far from solitary.

When you move to the country, you are starting with a clean slate and in a different social structure than in the city or suburbs. In an urban apartment building, you might rotate through many neighbors without ever knowing their names. But in the rural countryside, the folks around you often have been in their homes for generations. They've known each other all their lives. If you are lucky, you'll know them for a long, long time as well.

It is unlikely that any of your well-established neighbors will be moving anytime soon. With not many changing faces in the neighborhood, when someone new arrives, you become the hot topic of conversation, eliciting various possible reactions. These depend upon your behavior and whether you make the effort to be friendly and respectful to your neighbors. Reaching out to them can smooth over many potential rough spots.

The residents who have grown up in the area have an advantage that you do not. They understand the workings of the land, areas that get wet in spring and dry out first in summer. They know how the land has been used over the years and what crops may have thrived or failed. In addition to all of that local knowledge, these farmers can impart plenty of folklore. While it might be taken with a grain of salt, remember that most old wives' tales begin in some truth. For example, New England folklore says "Half your wood and half your hay you should have on Candlemas day"; in short, the measure for being ready for winter is gauging your necessary food and fuel carefully and accurately. Other adages include transplanting tomatoes into the garden on Memorial Day weekend and planting peas in the ground by tax day. These help to ensure a bumper crop, avoiding frost or vegetables ripening too late in the season.

At first, what can be tricky about neighbors is just how friendly they are. Especially if you come from the city, but even if you're just from a slightly more populated area, you might not be used to the locals dropping by unannounced just to say hello. This can happen when you're in the middle of a focused chore, when you might not welcome the interruption of a social call.

It may also surprise you that, along with your move away from "civilization," you can actually lose a lot of your anonymity. Yes, there are only a few households within a twenty-mile radius, but every one of those households knows your name and probably has an idea or opinion about your business. People in the country are not so focused on themselves and their own problems that they don't amuse each other with stories about the newcomers. This can be seen as busybody behavior, and sometimes it is, but it is also a comfort to know that folks care about you enough to notice changes on your land or in your behavior. Very likely, these neighbors will be the first to come to your aid if and when the time or circumstance call for it, particularly if you, too, have been friendly and obliging of their company and customs.

Interdependence

Rural farmers have long relied on local community members as a support group, and while your neighbors will be dropping by partly to size you up and see what you are doing with the property, they are also introducing themselves as potential help during a future winter storm or summer draught. Some may try to take advantage of you for a quick dollar for their work, but others will offer good advice and support throughout your time in the country.

Community Events

A town picnic, parade, or celebration provides the best opportunity to get acquainted with the local people. Getting out into your new community at such events can help build bonds with your new neighbors and, depending on their themes, can also be a way to find other rural farmers. You can feel isolated in your country home, but you also might be pleasantly surprised at just how much is going on around you.

You can also host community events, including tastings for your farm's produce.

The local town office may have postings for upcoming community events, or you can ask the neighbors who stop by about upcoming goings-on. Many small rural towns have a general store that is a hub of activity and serves as the central information center. It may sound a bit quaint and old-fashioned, but it is still the case that time spent at your general store will include locals discussing special news about events and community happenings.

Some city people are surprised to realize that the good old-fashioned country fairs still thrive as the main event of the year. They are wonderful entertainment and also a way for farmers to visit with each other and show off their goods at the end of a busy growing season.

Other Farmers

Neighbors can help you with local knowledge, and the town office can often provide a sense of community, but sometimes you really need the more specific help and support of other people doing this homesteading lifestyle that you are involved in.

Finding people who truly match your lifestyle and understand what you are doing doesn't have to be a challenge. When you first go rural, it may seem like you're the only person who understands what it is like to live without a flush toilet, care for animals every day, or live an hour removed from the nearest big-box store. Settle in for a while, and you will likely find that there are people nearby, if not your immediate neighbors, following the same lifestyle for exactly the same reasons that you are. They, too, know some of the same challenges that you face.

Meeting these like-minded folks requires some perseverance, because most of the time they're just as busy keeping up their homestead as you are. That is the blessing and the bane of rural living, that there is always something to do. At the same time, you and your family need to feel connected with other folks. As human beings, we crave meaningful relationships where we can share both our disappointments and our joys—and, in addition, we can benefit from hearing how others learned to cope with similar problems. Sort of like "many hands make light work," friends are almost never a detriment to our mental and emotional well-being; quite the opposite: friends are essential.

The internet can connect you with other farmers, sometimes with someone on a neighboring farm who has the same lifestyle. Other times you can connect when looking to buy or sell livestock or equipment. Often associations for farmers and homesteaders are organized in your area, some might be for young farmers, organic farmers, or off-grid folks. It can take a little bit of research, asking questions of the locals, or digging around on the internet, but after establishing yourself in the community, you will probably hear of other people in the area with similar lifestyles.

The Silence of the Country

You probably didn't think that the sound decibels of life would be an adjustment when you moved to the country. But the volume, especially at night, is noticeably turned down. Depending on how far you go into the wilderness, you may have little to no traffic noise and, for that matter, no light pollution. It is an idyllic existence under the stars, and this can also be a little bit unsettling at first.

A still, moonlit night on the farm.

If you aren't used to it, complete silence can be unnerving. Silence translates to every little snap and noise in the woods reverberating, and every creak of an old house amplified like ghosts. After several years of rural living, we check the window and watch every car that goes by our home after dark until their taillights disappear. Because it is so unusual to hear or see a car at night, you always wonder where they are going and feel the need to verify it isn't stopping at our address.

Alertness is a good thing. This sense of heightened awareness may mean you hear a predator before it gets to your chickens. But it is also important to be able to relax, and it often just takes time and getting used to what noises are normal versus what should be cause for attention.

The Quiet Night Life

It is also true that most country towns roll up the sidewalks after dark. Not because no social life exists, but such activities can be a little bit irregular. For farmers, there's livestock to care for, bring into the barn, and feed when darkness starts to fall. They know they'll have to be up at 4 or 5 in the morning to start chores over again.

If you are looking for an active night life, country living may not be for you. While plenty of steps can be taken to ensure you still have a busy social schedule, when it comes to going out for evening entertainment, you might find yourself with a challenge. Seeing your favorite bands may mean driving several hours to the nearest city, and you often know the staff of the few local restaurants by name.

For those that love the country life, the extra effort to get out can make the evenings on the town seem all the more special.

Emergency Services

What Goes in a Family First Aid Kit

A first aid kit can be extensive or basic, and might include herbs or homeopathic remedies. These are some key items you should never be without:

- Band-Aids of all sizes
- Adhesive cloth tape

- Gauze strips
- Antibiotic ointment
- Roller bandages
- Oral thermometer
- Tweezers
- Antiseptic wipes
- Aspirin and/or Ibuprofen
- Scissors
- Non-latex gloves
- Flashlight with working batteries
- Magnifying glass

Remember to keep the family first aid kit in a safe, easy-to-access space. Everyone should know exactly where it is, and emergency contact numbers should be posted nearby.

One of the most often overlooked and important aspects of the difference in country life is how long it can take you to get help in an emergency. Unfortunately, local fire departments are often volunteer run, and they can take longer to reach emergency situations than city departments do. Emergency crews have farther to travel, and depending on your location and your driveway, there might be trouble with readily accessing the situation. Fire and ambulance crews are both potentially delayed in the country.

This is a compelling reason to build a strong social network with your neighbors, who can usually get to you much faster than emergency services can. Friends and neighbors might be limited in how they can help, but you may be surprised at what they are able to offer in support.

In addition to true emergency services, power outages present another problem, especially if there's a serious delay in restoring your power. On-the-grid homesteads in rural areas may find they are off-the-grid for weeks, sometimes in extreme cold, especially in winter. If you require power for any reason, especially medical, a reliable generator is a worthwhile investment. Always have an emergency preparedness kit ready and a plan for food, water, and warmth in days without power.

What I have learned about country living is that, fundamentally, it is not as remote as it seems. They may not be the folks you expected to have as friends when you moved away from an urban setting, but there are neighbors who will watch out for you and stop by when they know something is not right. There are folks around the countryside who can offer you advice, and if you keep an open mind, you can find supporters everywhere who will lift your spirits and instill a feeling of "I can do this."

Further Resources

- MOFGA (Maine Organic Farmers & Gardener's Association): An organization in Maine that offers educational resources, workshops, an annual fair, tool lending programs, and ways to connect farmers with landowners. mofga.org
- NOF (Northeastern Organic Farming Association): An extensive program serving New England and the Tri-State area with chapters in seven states. Includes many resources for organic farmers. nofa.org, or do an online search for your state's chapter.
- Homegrown Organic Farms: An organization connecting family farms and organic farmers in California. hgofarms .com

- National Young Farmers Coalition: An exciting coalition focused on bringing more young people into farming and helping them succeed. youngfarmers.org
- World-Wide Opportunities on Organic Farms: An extensive international network that pairs up would-be farmers with farming opportunities, including assisting in establishing farms in remote and hard to access areas. wwoof.net
- Farm Aid: Not only a concert host, Farm Aid's focus is providing support and resources for family farms. farmaid.org
- The Livestock Conservancy: Dedicated to protecting and promoting heritage breeds of livestock, the Livestock Conservancy also provides resources for farms raising sustainable heritage breeds. livestockconservancy.org

Questions Before You Leap

- Are you prepared for helpful neighbors dropping by to chat? Do you have a plan for getting work done during a visit, without appearing rude?
- What will you do for social interaction?
- What resources does the nearest town or city have for entertainment?
- Are there other farms near your property?
- Do you desire a busy social schedule or night life?
- How prepared are you do to minor first aid in an emergency situation?
- Who can help you the quickest in an emergency situation?

CHAPTER 7

Seasonal Living

One of the true beauties of rural living is how close and connected to nature you become. It is no longer a hot season and a cold one: the nuances of each week as the planet moves through its phases are apparent and immediate. This powerful connection gives you a fresh perspective, just as it also presents unforeseen challenges. In fact, trying to keeping up with Mother Nature's changing and sometimes dramatic behavior is often the biggest obstacle for the modern homesteader.

Spring

Spring is an exciting time on the homestead. Everything seems possible and worth all of the long, cold nights of winter spent stoking fires and cracking ice out of water buckets. Some may not realize this, but spring starts making herself known as early as February. The days begin to lengthen, the slant of the light changes, and winter storms bring warm fronts instead of cold fronts. In many areas across the United States, sap will start rushing in the trees, signaling the first harvest of the new year for many country folks: maple syrup.

If you have sugar maples or rock maple trees on your property, maple sugaring is remarkably easy. It is possible to identify them in the winter, but it is easier to scout trees in the summer and tag them to indicate which ones to tap when the sap is running. Sugaring supplies are available at many country stores in early spring; all you need is a tap or spile, hammer, drill, and a bucket with a lid. Drill holes around chest height on the side of the tree that receives the most sun. The tap or spile will indicate how deep you should drill. Once your hole is drilled, hammer the spile in place and hang the bucket below. Check pails daily and collect the sweet maple sugar water.

Sugaring is done in early spring, often when there is still snow on the ground. Here in Maine, tapping for syrup usually starts before the end of February. The conditions for sugaring are days above freezing and nights that dip below freezing. Sugar water is delicious on its own, but bring it to a boil, then simmer, and

In spring the apple trees are in blossom, and everyone is excited to get back outside.

stir the pale sap until it is reduced into the dark, rich maple syrup that we love to pour over breakfast pancakes. It can take up to fifty quarts of sugar water to make a quart of maple syrup, so make sure you've gathered plenty before you start reducing.

Sugaring season ends when the nights remain above freezing, but spring is only beginning. If you've spent your winter well, you will have made plans for the summer's garden beds. Once the earth is workable, you can till or cultivate the garden beds, mixing in fresh manure or compost as needed. Most seeds shouldn't be started outdoors until the soil temperatures are warmer, but with brightly lit south-facing windows, you can get a head start on summer by propagating seeds inside using seedling trays. If you are eager to get into the soil, cultivate inside your greenhouse, hoop house, or, if you have one, a simple cold frame. With a cold frame or greenhouse, you should be able to harvest some hardy vegetables throughout winter, and certainly by the time the nights are above freezing.

For those keeping animals, spring is a feverish time filled with the welcome excitement of newborns. Those raising chickens should see new chicks arriving in the mail in early spring or be setting their incubators up. New chicks need a space prepared before they arrive or hatch: a brooder box with a lamp keeping it around 90 degrees and fresh food, water, and bedding. When your baby chicks arrive, it can be hard to believe they'll develop into full-grown chickens, but take advantage of time and start building their adult housing now. You will find it is better to have housing ready than to be suddenly scrambling at the last minute, just when there's already a long to-do list. As busy as spring is, you'll have more time now than you will in the active rush of the summer months.

Goats, sheep, and cows all may have their young in springtime as well. Watch for signs of labor and be prepared for deliveries with a first aid kit and your veterinarian on speed dial. While most deliveries go without a hitch, it is good to have done some research and be prepared for any eventuality. Kids and lambs will bring an uplift of joy, bouncing around the homestead right away. Animal mothers often raise their babies without any human interference; however, you may have to teach a newborn kid how to nurse or resort to bottle feeding a rejected lamb.

Our first year raising goats, one doe gave birth to triplets, with the third weighing under two pounds (small, even for a Nigerian Dwarf). Sadly, the mother rejected him. In this situation, it is important to milk the mom for the first 24 hours to ensure that your kid gets enough of her colostrum, and after that you can use formula specific for the species. Feeding a bottle baby is a re-

Our goat, Lucky, the first bottle baby on our farm.

markable task, and it is also a huge responsibility, not necessarily something you should wish for. The life of a bottle-fed kid is entirely in your hands, which is a gift and a demanding chore. Newborn goats need to feed every four hours, even through the night.

There is also the possibility of the mother animal needing assistance, and other complications. While healthy, happy, bouncing kids that need little attention except for cuddles is the dream for every farmer, keep in mind that any time you breed an animal it's possible you'll be spending weeks tending to the mother or her kids for a large part of the day.

Usually we breed animals in order to milk them. So if you are preparing for kidding or lambing, you are also setting up a milking station and getting ready to milk. A basic milking stand for a goat is easy to make following plans available in goat care books or online. You will have to invest in a way to store the milk that is safe and sanitary, even more so if you plan to sell your dairy goods to the public. This can be a simple stainless steel setup with a fridge. To make cheese and other products, you will need specific pieces of equipment. Everything will have to be kept clean and sterile in order to meet state health and food standards. Spring is the perfect time to review your dairy supplies. Give the equipment a good cleaning and purchase any new items for your dairy operation before milking starts in earnest.

Winter is hard on equipment and buildings. Months of buffeting wind, drifts of snow, and expanding and contracting from freezing and thawing will inevitably cause some breakages, leaks, or cracks. Often during the depths of winter, the solution is to tape it over, tie it up, screw a board in place, or otherwise use a carpentry Band-Aid. Now when the weather is tolerable is the best time to create more permanent solutions to any fencing or housing issues.

Take the time one spring day to walk your full fence lines and do any necessary repairs, including replacing gates and fencing. Some may be costly fixes, but they will keep your animals safe and out of your gardens and crops. Check your buildings as well and make sure windows are well in place, doors open and close properly, and so on. It is a good idea to just give everything one thorough once-over.

In addition to your structures, check all equipment, paying particular attention to the pieces you'll be using during the summer. For the modern homesteader that may include tractors and rototillers, chainsaws and lawnmowers. Even if you have no motorized equipment, you still need to check your essential tools. It is much easier to fix broken pieces and replace parts before you actually need the items.

If you do have a tractor, spring is a good time to have it checked by a mechanic and to change the oil and fuel in your other machines. Rototillers, especially hand-operated ones, are notoriously finicky, and many gardeners are grateful they only have to be running a couple of times a year. Check the air filter on a tiller, and also make sure the tines are in good shape and sharp, otherwise you'll get poor performance out of your machine.

Do not forget to check the edges on axes, handsaws, hoes, shovels, and pitchforks. Sharpen if necessary with either a stone or file. They do not need to be razor sharp (except the axe) but honed enough to cut through sod and roots without resistance. If you spend a day or a week sharpening garden tools and handsaws, chances are you'll be glad you did when the busy summer arrives. Also check handle seatings—there is nothing worse than a handle coming loose from an otherwise reliable tool while you are using it—and tighten or replace when necessary.

With all this checking and rechecking and repairing, you don't have an overflow of extra time when true spring arrives full on. Take advantage of the earliest spring season to organize storage spaces. Your freezer should be emptied out of many goods following a long winter, and maybe it is time to clean and defrost it and organize the contents by date. Go through your pantry and toss out anything that has expired. When you are in the midst of a canning or preserving frenzy next August, you won't want to have to search for the right spice or fret over storage space.

You can do similar organization as you put away your repaired and sharpened tools, making sure they're lined up in an easily accessible place. Any area used for potting or transplanting seedlings should be ready for the job; review and replace any broken or cracked pots before the season starts in earnest.

All of this may seem like extra work that does not immediately reward you with some produce or goods. But being prepared will make bringing in the harvest that much easier, and sometimes be the difference between success and failure.

While not critical to spring on the homestead, the warming weather can be a great opportunity to purchase any replacement winter equipment. Whether you are looking for a larger woodstove, a snowplow, an upgrade to a propane heater, spring might be the best time to look for sale prices. When fall rolls around again, everyone else will be searching for winter equipment and the prices will be higher.

Summer

Summer is the season we have been waiting for: those long, warm days and extra hours of direct sunlight, when everything swings into place. Starting with spring peas, radishes, and greens, your garden will produce fresh vegetables as summer begins. Nature's own garden is abundant with forageable goodness, from elderberry blossoms to dandelions, fiddleheads, and wild berries. In highly seasonal areas such as the Northeast, the forests will come alive with birdsong and the sounds of creatures that have either hibernated all winter or are returning from their winter homes.

And of course, on a homestead, summer is your make-or-break season. You may not be harvesting every day at the height of summer (much of that will happen in the fall), but everything you are doing now points to the goals that you have made to increase your harvests and will determine how busy your fall is and how much produce you will be able to put away before the winter months.

The garden is the primary focus on most homesteads in summer, when all that winter planning and spring seeding pays off.

The garden in full summertime ripeness.

A garden does not just grow and give you fruits, it needs to be cared for every step of the way. If you want to gather a good harvest, start with caring for the soil early in the spring, and continue to weed and water all summer. Weeding, which gives your plants room to grow and eliminates others competing for nutrients, often has to happen weekly. In a limited garden space, consider companion planting, sowing different fruits or vegetables next to each other that will benefit from each other's nutrients to grow. Research before throwing various plants in the ground next to each other, as it all has to do with the types of nutrients each likes. However, beyond companion plantings, make sure that your vegetables all have ample room and adequate air circulation. In a weedy garden, plants have to fight with a population of other hungry ones for the nutrition they need to grow, resulting in fewer, smaller fruits and a disappointing harvest.

A garden also needs water. If you do nothing else for your garden all summer, make sure it is watered. Some plants prefer more water than others, but all will be thirsty during the height of hot summer days. With a quarter-acre homestead garden, I am often outside up to two hours every evening watering each plant by hand. Drip irrigation systems work well but usually has significant up-front investment in the right materials. Daily time spent with the plants in your garden keeps you in personal contact, aware of the exact growing conditions. A sprinkler system wastes water and does not get nutrition directly to the roots of your plants. The best way to water is directly with a hose or watering can, plant by plant.

If your water source is some distance from your plants, you can fill a livestock trough and truck it to your garden site, then use a pump to work a hose from that. You can also fill up multiple five-gallon buckets and truck or carry these to the beds. Ideally you have a nearby water source, since the garden will be your single biggest consumer all summer.

Your garden also needs some fertilization, depending on the health of the soil. Review the plants' stage of growth, and have your soil tested to know what you're working with. In-depth soil tests are usually available through state agricultural labs for a small fee, and they are well worth the cost.

And then there is the reason that you keep your garden: the harvest. Harvest times and quantities depend on what is growing; suffice to say that much of the summer will be spent picking, eating, and preserving your produce. As often as possible, you will be eating fresh off the vine, and it's very likely you'll be trying to sneak some zucchinis or cucumbers onto neighbors' porches.

Vegetable and fruit harvests can be preserved in a number of ways. Research specific recommendations for each type of produce. The most common ways include pickling, fermenting, canning, and freezing. Another option is storing vegetables in a root cellar or the corner of a cool, dark basement.

Pickling isn't just for common cucumbers; you can pickle practically anything. Very generally, pickling is preserving food in vinegar to prevent spoilage. It requires a fair amount of effort in the kitchen, starting with washing and chopping the vegetables and sterilizing jars and lids by boiling. Stored in your cupboard or pantry, these will last all winter long, while "quick pickles" made without boiling and stored in a refrigerator last about a month.

Freshly fermented beets will last us all winter.

Fermenting is very similar to pickling but uses a salt brine instead of vinegar. It encourages the growth of good bacteria, which will help to prevent food from spoiling, and it also lends a zesty flavor to most vegetables. Fermented vegetable flavors may change over time, but properly handled, they can be tasty for years.

Canning is the most time-consuming way to preserve vegetables but also one of the most effective. It involves filling a sterilized jar with food, sometimes raw and sometimes blanched, and placing the sealed jar in boiling water for a certain amount of time. When you remove the jar from the water, the air leaves it, creating a tight seal. Like fermented goods, canned goods can last you several seasons.

Freezing is another great way to preserve food. I would recommend a large chest freezer as a key appliance for any on-the-grid homesteader. You can freeze meat, vegetables, fruit, even eggs and cheese. As long as it does not thaw, it can last years. Some foods freeze better under certain conditions than others; for example, zucchini should be shredded before freezing. Others can just be tossed in the freezer in plastic freezer bags or containers and enjoyed when you are ready.

You can also try dehydrating certain vegetables and experiment with other kinds of preserving when it comes to meats and dairy products.

If you keep dairy animals, you should be at the height of milking in the summer months. Remember that this involves not just the daily predawn task of milking, but also the preservation of the milk and hopefully the creation of dairy products. From cheese to caramels to ice creams to soaps, milk can be used for many tasks beyond flavoring your coffee. With this livestock, your summer will be busy with both milking and making.

Seasonal meat animals, such as broiler chickens or hogs, arrive as young hatchlings or piglets in spring, but much of the time caring for them is during the summer. For delicious, high-quality meat unlike anything from the grocery store, rotate their pasture space. This allows them to forage for fresh, natural food every day and minimizes how much grain they consume. It does, however, mean you'll spend time each day checking their pastures and moving them to new fields if necessary.

In a seasonal climate such as in Maine, summer is about the only time for any large building or expansion projects. It's good to have any work well planned out in advance, so when the good weather arrives, you can make progress and get the job done before hard frost returns. Major clearing of brush or cultivating of new fields happens in the summer, as well as putting in fences or outbuildings. If you have a big project that must be completed before cold weather sets in, get it organized and plan to accomplish it in summer.

Some summer jobs can be accomplished yourself, others require construction crews or specialists. Be prepared to arrange for such help, since folks who pour foundations or dig septic systems are just as busy in summertime as you are.

If you harvest your own firewood, you will want next winter's wood cut and chopped in early summer. If you cannot cut wood until fall, plan to use that wood the following winter and not the coming immediate season. Wood harvest starts with felling large trees, limbing them, sawing logs down to size, and finally chopping wood into stove-sized chunks. This can all be done yourself with the right equipment, but it takes time, so plan on spending significant hours in spring and summer on this job. Once wood is cut, it needs to be properly stacked (preferably inside a shelter or under a lean-to) to dry for about six months. Wet wood doesn't just burn poorly, it's dangerous to use because of the potential of creosote buildup.

During your first couple of seasons on the homestead, you might feel like you are behind and constantly trying to catch up. Stick with it, and you will soon get the timing of how to put up good supplies of firewood for when you need the heat. In the meantime, do not feel ashamed to call a local supplier and get precut wood so you can be warm all winter long.

Those who hay their own fields will also have a busy summer. It's a good idea to have them mown once or twice a year even if you don't use the grass for hay. Sometimes a local farmer will do it for you and accept the hay as payment, but if you have a tractor or even a hand scythe, you can keep fields under control yourself. It is necessary to do this in order to keep your fields clean. Left completely to nature, alders and small weed trees will start springing up within a summer or two, and your open field will quickly become puckerbush.

With actual hay fields, most take two cuts and some are able to get three. The difference between first-cut and second-cut hay is what plants are in bloom at the time of harvest. First-cut hay is mostly grass without blossoms, usually harvested in June. It is both coarser in texture and also lower in nutrition. Cows and sheep eat primarily first-cut hay because they do not require the heavier nutrients available in second-cut, and because they simply are not picky eaters. Horses should also be fed first-cut, they are unable to digest richer second-cut hay.

Second-cut hay, harvested in late July or early August and occasionally again in September, is sweeter and rich in nutrition and calories. It's full of alfalfa and legume blossoms, great for the digestive systems of goats, and serves as a good special treat for other grazers. Generally speaking, animals go wild for second-cut, which is also more expensive. So, most farmers feed first-cut primarily and second-cut to supplement grain.

Whatever the cut, hay is made on dry, sunny summer days. Hay is turned or "tedded" a few times in the field before being put in windrows for baling. Putting away wet hay is a waste of time and energy, as you may as well be putting away trash. It will spoil and mold and be inedible. Anyone who has ever spent time on a farm can recall the sweaty, hot days of haying where little bits of straw get stuck in your clothes and sweat trickles down your back.

If you do not cut your own hay, look for it locally and purchase enough to get through winter, depending on how many animals you keep. Buying in winter is expensive and impractical, so borrow a trailer if you do not have one and load it up with everything that you need. Store your hay in a cool, dry place—dry being of paramount importance.

Just because it is summer also does not relieve you of planting seeds. While your spring seedlings should now be producing heavily and keeping you busy in the kitchen, it is time to get any fall crops in the ground. Second crops of beans and peas can be planted, and brassicas and root crops will thrive when the cooler fall weather comes. Many even taste better after a light frost.

Any new animals that you added to the farm in the spring will now be ready for adult housing. Their integration with the flock's or herd's existing social structure can be stressful and requires careful watching. If new poultry or animals are from another farm, quarantine them first to ensure that they are healthy. Animals raised on your own farm, from your own stock, will likely fit in almost immediately, as they will have their mothers to protect and introduce them. For grown animals integrating with an existing herd, start by allowing them space next to, but secure from, your existing animals and slowly let the two groups grow used to each other.

While most of summer is a busy time on the homestead, it is important to take a break when you can. The beauty of summer should not be overlooked. Try to take a few moments at the end of a day to sit back and enjoy the sunshine.

Fall

With the start of shorter, cooler days comes the urgency of winter preparation. If summer time is about preparing for the distant winter months that seem almost imagined, then fall brings the insistence of knowing winter is directly ahead.

Before the weather starts getting chilly, you should have your stove, fireplace, chimney, and any other heating devices checked. A chimney should be inspected by a professional sweep and cleaned if necessary. Using a professional means paying a fee, but

Fall brings foliage along with cooler weather.

it is something that brings peace of mind concerning any unfortunate consequences of an unclean or cracked chimney.

Your woodstove or fireplace should be thoroughly cleaned and any parts that need replacing should be addressed. Check the sealing and replace if necessary. Stove sealing is inexpensive and easy to apply yourself. Check that dampers or flues open and close properly and inspect the top of your chimney for birds' nests and rodents.

Non-wood heat sources should be checked as well. Start up any heater and ensure it's working properly while it is still warm out, instead of having it break in the depths of a cold snap. For any heater that runs on fuel, be it propane, kerosene, or gasoline, have the fuel checked and refilled so it is ready to go.

Canning and preserving the food from your garden kicks into overdrive as more vegetables ripen. In addition to crops such as tomatoes, potatoes, onions, brussels sprouts, winter squash, cabbage, and the ubiquitous pumpkin, fruit trees also are ready to harvest in the fall. Scores of people head to local orchards to pick their own apples as a family event. For homesteaders, it is an opportunity to put more food away for winter.

Like vegetables, many fruits can be preserved by canning. In addition to enjoying the juicy sweetness of canned fruits and juices in winter, you can also transform many into jams and jellies that last for months and preserve the taste of summer. Apples are particularly easy to preserve; many varieties can be laid out on a rack with open slats, or wrapped in single sheets of newspaper to remain dry and crisp well into the winter months.

Soil and bed preparation should be completed the season prior to planting. This time of year, while you are harvesting, you can also put garden rows to bed. Once a plant has finished producing, it can be uprooted and added to your compost pile. You can also leave some plants to rot in the garden, giving their nutrients back to the land for next year's plants to consume. This technique spares you some work, but some plants' large root balls will be a problem when spring planting comes around, and rotting plants can harbor bugs and disease over winter. Any overgrown areas that you've let go to weed should be cut back and cultivated or forked over quickly to lessen the amount of work you will have come spring. Plants that you are overwintering need to be bedded deep with hay, straw, shavings, or shredded newspaper so that their roots don't suffer even in the coldest weather.

You can keep on planting once the weather gets cold. Garlic does best if sown in October, at least here in Maine. Cold frames and greenhouses can be stocked with fresh greens to get you through the winter weather. Spinach, leeks, carrots, and kale can all thrive in a simple cold frame, even when there's several feet of snow on the ground around them.

Not to be overlooked in winter preparation are your animal shelters and pastures. Winter is hard on fencing, so check once again for weak spots and repair as necessary. Also ensure that gates and doors open and close easily. To keep most livestock, including poultry, warm in winter, the key is to provide them a shelter that protects against wind. Regardless of the animal, everyone wants to be able to get out of the wind and also breathe easily with good ventilation. This is best supplied by overhead vents or cupolas, some kind of opening above their sleeping area, that allow moist air created from steaming water buckets, breathing, and droppings to escape so that it isn't captured and turned to ice on your animals' bedding.

Preventing wind from blasting through your stalls or shelters means screwing or nailing boards over any large gaps, shutting windows, and sometimes covering them with plastic if the trim is not sufficient. When you can block a hole with scrap wood, it's also possible to nail or staple up a thick sheet of cloth, such as burlap or canvas.

Before winter comes, give your livestock shelters a really good clean right down to the floorboards. During winter, it is common to use a "deep litter" method, allowing soiled shavings to build up underneath new, fresh bedding. This keeps a nice thick layer between your animals and the cold barn floor, but it's good to start deep litter from bare earth or floorboards so it doesn't build up to quickly. Cleaning stalls out entirely during freezing temperatures is almost impossible, as droppings and urine freeze the bedding to the floor. During the coldest days, just add more shavings on top to keep your animals happy. Make sure to stock up with bedding so you'll always have a few extra bales of shavings or straw when a storm hits.

Like heating devices, all other winter equipment should be started up and tested before they're needed. Tractors should be running smoothly, a snowblower if you have one should be ready to go, and a collection of snow shovels should be at the ready for a storm. If you use heated water bowls for your animals, make sure they are working.

Summertime equipment needs maintenance also. Before putting away your lawnmower and rototiller, stabilize them so they survive winter in good condition. Empty fuel from any small engine equipment before being put away; otherwise the gasoline will break down and can destroy your engine. If your tanks are not drained, use a fuel stabilizer to avoid any unpleasant spring surprises.

It is a good idea to do other equipment checks now, such as replacing spark plugs that may have worn out over the summer. Change the oil and replace air filters either now or in the spring before use and give everything a thorough cleaning. Clean and oil handsaws and shovels so they're dirt free and the debris from summer use does not corrode their materials over the winter. Finally, put all of your equipment in one area so that you can find it easily and get it started up again when spring arrives.

Just as important as winterizing your barn and pastures is winterizing your house. Being cozy during the long winter months has uncountable benefits.

As with your livestock's shelters, first seal any spots where wind can get into your house. Doors and windows are the most obvious places where you can lose heat, but don't forget to check cracks between floorboards that might be direct lines to chilly basement air. Windows should be sealed over with plastic. Today there is plastic available that, using special double-sided tape and a hair dryer, provide you with a clear view outside while still keeping out the wind. Thick curtains will help keep the wind at bay on especially cold days.

Winter

On the homestead, wintertime is only slightly less busy than summer. It is a different kind of busy, mostly structured around planning and organizing. It's also a time to rest your bones and exercise your mind.

Look back over the past season's accomplishments and assess how to make improvements in the future. Let yourself be inspired with new ideas and project planning. Planning the next year's garden means studying the successes and failures in the past. Gardening is not something to back down from, it is a practice that continually improves season after season. Maybe you didn't get the results you were hoping for, maybe you battled drought or bugs, or you didn't get something in the ground early enough and its produce was not ripe before the frost. Maybe one plant was decimated by a pest. Try again. Be assured that your knowledge and experience build with each year.

The first step toward planning your next garden is to decide what you'll grow. Are there new varieties that you want to try? Some that you want to omit? Once you've narrowed your choices down, you can start filling out garden plans. Everyone has their own style; some like using graph paper and colored pencils, at the very least you should draw an outline of your garden with plants filled in to give you an idea where everything goes. This way you can decide on companion planting, accommodate for plants that take up a lot of space, and place certain crops in specified areas, like shade-tolerant ones in one end of the garden and those that prefer direct sun in the other.

You can make several versions of your garden plan, rotating plants out and adding replacements as the seasons change. Good

The snow piles high in wintertime.

planning will make spring planting chores easier, but allow yourself to remain flexible because no matter what you put down on paper, some things are bound to change.

Armed with your garden plan, you can start ordering seeds. By February, many farmers have seedlings sprouting under grow lights or in bright, south-facing windows of their homes, already getting a start on the growing season. Browsing through seed catalogs is a favorite part of winter, supplying the necessary promise that new growth is bound to happen and instilling the auspicious knowledge that the upcoming season will surpass the previous one.

Winter is also the time to plan for any animal purchases you hope to make in the coming spring. My personal favorite way to spend a snowstorm is curled up by the fire with a hatchery catalog in hand, envisioning the coming year's flock. Chickens, ducks, and geese may be mail-ordered, so you can make up lists of your ideal birds, figuring out what qualities you are looking for, and then place your order and wait for spring.

With livestock animals you generally do not order through a mail order catalog, but you can contact breeders and local farms or start getting a birthing stall ready. Animals are bred in the fall or early winter. Breeders know their pregnant stock needs extra protein and vitamins during the long winter months to ensure the health of their developing young. If you have pregnant livestock, you'll be delivering babies by spring, so research what is required and all of the eventualities, and make sure to have a clean, warm stall ready to go.

As with a garden, the failures and successes of the previous year should figure into your animal planning. If you have found a niche with a particular type of livestock, it might be a good idea to expand your stock. For example, you might find yourself ordering a dozen new ducklings because your duck eggs turned out to be a more popular farm stand staple than your chicken eggs. If you struggled with moving a chicken tractor and processing pastured meat birds, you may want to think about keeping them in a different setting or taking a year off from raising them.

And there are always the new animals you want to try out. If you have always dreamed of keeping goats or raising pigs, winter is the time to start researching in earnest, making sure that your land and budget can support this new animal venture, planning how you will house it and what you will do when it comes time to butcher or start milking. With any new livestock animal, remember that this creature's life is in your hands, so before you bring anything home, read a few books on the subject and make sure you are prepared.

Winter is a time of rest, but sometimes resting can mean work on the homestead. The woodstove will need tending all day and through the night. On the coldest nights, I have been known to set an alarm to get up and stoke the fire. Bringing in firewood will keep you busy as well, so plan to bring in a certain amount daily, which will create a nice buildup of dry wood for the days you might miss. While our stove goes almost constantly, we do make a point to let it burn down once a week in order to clean out the ashes and keep the stove working safely and efficiently.

A lot of winter's wear on buildings and equipment waits until spring for repair. Some issues (namely anything involved in keeping you warm that breaks) won't wait, and others may only need a Band-Aid. Winter is tough on everything, especially anything mechanical. On days warm enough for your fingers to work, you will often find yourself in the barn tinkering with a piece of equipment, or out in the field chipping ice so that the gates still close completely.

Clearing snow keeps us fit all winter long.

Snow removal is a huge part of winter in any area that faces significant accumulation. After a storm, you will be shoveling for a few hours to get everything accessible again. Many animals prefer or need a space shoveled out for them to exercise, so that is extra work if you have any livestock. With a tractor or snowblower, the removal is easier, but still time-consuming. If you have several storms in quick succession, there will be the problem of where to put all the snow. From the start, take the time to make your paths wide and to clear large spaces; the extra effort might pay off—or else you can find yourself enclosed with abundant snow accumulation after a few more storms.

Snow can make winter long, with plenty of demanding work, but it is the cold itself that can be dangerous. Always make sure your animals have warm water to drink, plenty of food to keep their systems going, and warm dry bedding. Even on the coldest days, ensure that they are as comfortable as possible.

You will find that "winterization" continues through the season, as cracks you didn't know were there appear when the wind howls. Keep plugging holes or seal off a door that you aren't dependent upon. It always gets colder than you thought it would when you were winterizing in your t-shirt back in September.

The most important part of winter is enjoying the fruits of your summer harvesting and canning, and getting some rest. It's the only time of year that might provide you with a few days of actual break time, allowing you to recharge your engines and read up on how to improve your gardens and livestock. The simple act of cooking up a meal comprised of home-raised produce in the middle of a January snowstorm can make everything worthwhile.

Questions Before You Leap

Spring

- What do you plan to harvest first in spring?
- Do you have maple sugaring supplies?
- Do you have a birthing kit for any expecting animals?
- Do you have a brooder arrangement for new birds arriving?
- Are you set up to safely and sanitarily milk your dairy animals and process the dairy?
- What materials have been damaged over the winter?
- How are your fence lines?
- Is all equipment ready to go for spring and summer work?

Summer

- What can you forage from the wild or harvest from your garden during summer?
- What crops are going into the ground?
- Do you have a plan in place to keep up with weeds and keep the garden watered and fed during the busy summer months?
- How will you be preserving your crops?
- Will you be butchering any animals this summer?
- Do you have firewood cut and stacked for the coming winter?
- What is your summer plan for clearing any overgrown or weedy areas on the farm?
- Will you be haying your fields?

Fall

- Is your woodstove and chimney ready for use?
- Are any other non-wood heat options running smoothly?
- What's the plan for putting your gardens to bed and mulching them over?
- Do you have any cold crops going in the ground in the fall?
- How weather-tight is your animal's housing?
- Are winter machines such as snowplows or snowblowers running smoothly? Do you have enough shovels?

Winter

- Are your animals being bred for spring kids/lambs/calves?
- What could you improve on last year's garden or crops?
- What animals are you looking to add to your homestead come spring?
- Where will you be putting all that snowfall?

CHAPTER 8

Raising and Educating Children on the Homestead

by Julie Letowski of folk-ware.com
and Instagram "Homesweethomestead"

Years ago when my husband and I were living in the city, it was in part the thought of raising children in an urban environment that led us to explore a more rural path. While cities have so much to offer families, for us there was an undeniable tug to give our children something else. We wanted to afford them the rich experiences offered by land to roam, animals to love, and a garden to tend. Since those first bucolic dreams began percolating in our city-weary minds, we've learned firsthand the ins and outs and ups and downs of raising a child while building a homestead. Bless the good earth, it's the best decision we ever made as parents!

When Is the Best Time to Start a Family on the Homestead?

You're not the first person to ponder this question, and you certainly won't be the last. In addition to many heart-bursting, wonderful things, having children is completely world upending. Deciding to add a baby to the demanding life of a homestead can be overwhelming, and many couples wonder if they can time this in a way to make the transition smoother. While a baby born in the fall would afford the new family time to acclimate in slower

Credit: Julie Letowski

Raising children can be a rewarding part of homestead life.

winter months, mama would be her most pregnant during the unforgiving months of summer. Conversely, a baby's arrival timed with the appearance of rhubarb does not exactly bode well for getting the garden off on its best foot. The truth is, there are great arguments to be had for babies born at any time of the year on the homestead, so much so that I've come to believe it's less about seasonal timing and more about where you are in the journey of establishing your homestead. While anything is possible and plenty of homesteading families make a wide array of scenarios work for them, if you have the option, welcoming babies to the homestead is much easier after having spent a few seasons laying its groundwork. It can be said for both homesteading and babies: the work required is far more than you will ever imagine prior to taking on the job. Don't do yourself the disservice of underestimating either. That said, they're both the most fulfilling and rewarding ways to spend one's time, and I highly recommend you do both, in tandem eventually, should you have the interest.

Build a Homestead Apothecary

While it isn't always the case, many homesteads find themselves a little off the beaten path. It's one of their greatest attributes, in my opinion! That said, it can put treatment for the minor bumps and bruises of childhood at the end of a long drive. Stocking your medicine cabinet in times of health will go a long way to ease your worry in times of sickness. As every parent knows, children spike fevers, take nasty falls, or develop any number of non-life-threatening maladies on their own schedule. It won't always be when the small local pharmacy is open. Consider circumstances with your children that wouldn't warrant an immediate trip to the doctor or hospital but would require some attention. What would make them feel more comfortable? What would make you feel more comfortable? Preparedness is the name of the game here and is as simple as taking a list of necessary items with you on your next trip into town.

In addition to stocking up on items from the pharmacy, you can take stock of what your homestead has to offer. Of all the skills that are developed on the homestead (and there are quite a few!), building a relationship with plants that can heal your children is quite powerful. It's not the path for every family, but for those who choose this, it's an incredibly valuable tool in the parenting belt. From batches of homemade elderberry syrup for immune support to soothing calendula salve for scrapes, there are ways to support the health and well-being of your children without ever leaving the homestead. Within many rural communities that have experienced a resurgence of young homesteading families, classes are often offered for parents looking to support their children's health via herbalism and the local plant life. There is also no shortage of online courses and literature to help get you started.

Parenting and Homesteading in Tandem

When our family pulled our U-Haul into the driveway of our 1850s farmhouse years ago, mid-blizzard, we did so with an incredibly active and curious two-and-half-year-old in the backseat. Our land hadn't been cultivated in decades, the barn was full to the rafters with junk, and given the work that lay ahead of our first growing season, there was an obvious shortage of hours in the day. In the early days, we learned how to parent and homestead just as much by finding what worked as much as what didn't. Through the ups and downs, I found that specialization, time block organization, and flexibility have had the largest positive impact on managing a homestead while raising children.

Credit: Julie Letowski

Kids are happy to help with berry picking and get a tasty snack in the process.

Specialization

For every endeavor on the homestead, there is a mile-long list of important details to go along with it. And while possible, it's not necessarily practical to take on all the information pertaining to all your homestead's moving parts, especially when raising children. These little sweethearts take up a great deal of brain space! If you are parenting with a partner, I highly recommend specializing. Specializing is a way to effectively manage the responsibilities of the homestead while not overloading your brain with too many choices and mountains of information.

To start, sit down with your partner and make a list of all the homestead's animals and projects. Here is an example of what that list might look like:

- Cows
- Pigs
- Chickens/Geese/Turkeys
- Sheep
- Garden
- Sugaring
- Milling lumber
- Preserving
- Establishing the asparagus patch
- Blueberry field maintenance

Determine what each of you are drawn to or have an aptitude for and divide the list evenly. While the weight of the homestead is shared evenly between partners, you now both have areas in which you can focus and develop your expertise, and others in which you can take a mental break. This is an example of what my husband loves to call working smarter, not harder. To more clearly explain, I'll share an example of how this works for us. On our homestead, my husband has always taken the reins with the cattle while I take charge of the sheep. We are a team as far as the work required goes—rotating their paddocks and doing chores together—but as far as what practices are implemented, knowing the ins and outs of the specific breeds we're raising, and so on, that falls to an individual. Not only does this give you an opportunity to focus, it saves a great deal of time that can be lost to back and forth conversations and decision making when you have two barnyard bosses researching the same thing.

Time Block Organization

Once you have specialized on your homestead, it's time to get to work. Ah! But what about the kiddo? Here's the deal: it may appear that our kids are living in our homesteading world, but really, we're just homesteading in our kids' world. But don't get discouraged! There are plenty of activities around the homestead that they can be a part of, which is especially true as they get older. Parents of tender young babes have to be gentle on themselves though because you can only accomplish so much in a day. There are meals, naps, moods, and more that make just having the kids "along for the ride" complicated. Take heart. It gets much, much easier and in the end will be beyond worth it.

One of the more useful tools we have implemented on our homestead is time blocking. This entails one parent being solely responsible for the babes of the house, while the other takes on homestead tasks with undivided focus. In times of intensely demanding parenting, time blocks have allowed each of us opportunities to accomplish important and necessary tasks while making sure our son had what he needed. Parents can swap roles from day to day or divide days up into shifts. One of the greatest benefits that has come from time blocking for our family is that we are able to be more present in whatever we're doing, whether that be taking our son for a hike or ripping shingles from the side of the barn, and because of that, we've seen a huge decrease in stressful moments where we're trying to do too many things at once.

However, it's worth mentioning that one of the greatest parts of growing up on a homestead is being immersed in its activities. Time blocking isn't a suggestion to remove kids from the goings-on but rather a way to manage the many moving parts of homesteading and parenthood with focus and minimal stress, especially when children are quite young and require a good deal of attention.

Flexibility

Much like life on the homestead, parenting is full of uncontrollable forces. One of the greatest gifts you can give yourself and your child is the permission to be flexible. While there will be situations that arise that need immediate attention —for instance, an animal out of its fencing—the bulk of what must be done can always be done later, whether that later be in ten minutes after you've soothed an upset child or in a few hours when your partner is able to empty your arms of children. There will be times when you finish your day feeling superhuman after having kept everyone alive, partially clothed, and well-fed, in addition to weeding the garden or rotating all of animals on pasture. There will be other days when you'll have to settle for far less. Flexibility is how you find peace with either outcome.

Surviving the Market

Many families find themselves in fairs, shows, or farmers markets selling the goods of the homestead, and many do it with children in tow. I can speak from experience that it's both rewarding and incredibly demanding. There are a few things you can do to help with the latter. First, before the market season starts up, request that your booth be positioned near green space if there is any. Our first market was in a library parking lot, and our booth was in the last slot, right next to soft green grass and big shady trees. Not only did it give our son a nice place to play, it also kept us away from the parking lot traffic. On days when I ran our booth solo, it was an absolute lifesaver.

In addition to getting an appropriate spot for your family at the market, bringing a little bag of tricks is essential. While on the homestead exploration and wild free-play is great, it can be overwhelming in a bustling farmers market setting. Mama and papa will feel much better, and do their market jobs more effectively, if the kiddos are nearby as much as possible. Back in our market days, I would pack more food and drink than seemed reasonable for a tiny person to consume and plenty of clothes for all kinds of weather and accidents. I would also bring special toys or activities whose newness alone would grab my son's attention for a solid chunk of time: homemade playdough, blocks that only came out on market days, or a bunch of animal figurines frozen in a block of ice and ready for excavation! Don't waste your money on new toys, just be creative!

Raising Little Homesteaders

As was certainly the case in my family, a great deal of people find that their journey back to the land is spurred by how they hope to raise their children. Open spaces, fresh air, and a break from a culture that oftentimes prizes media and consumerism over homegrown and handmade is at the forefront of many current and future parents' minds. Here are great ways to set up your children to be successful little homesteaders in their own right.

Start Them Young!

If your children are born on the homestead or move there while relatively young, baby-wearing is a fantastic tool. Slings, wraps, and carriers not only offer young babes the comfort of having their parent nearby, they also afford parents two free hands and an excellent view of that natural world that will surely keep the young one entertained.

Meet Your Child Where They Are

Children love to help. If, from very early on, we provide them with opportunities to actively engage with their environment, they will not only grow up feeling capable but also like the essential part of the homestead team that they are. When considering how to involve your children, let the work fit the child. Young children are particularly fond of scooping and playing in dried beans. I'm sure you've seen many home activity tables set up with a wide array of spoons, scoops, lentils, and beans. This interest translates perfectly to scooping grain for animals. Initially, they may not have perfect aim with the bucket, and perhaps you'll have to take over after they lose interest, but it's not about being a blue-ribbon chicken feed scooper. It's about involving your child with the rhythms that are essential to your homestead's success.

Let Your Child Take the Wheel Hoe!

I have never met a child that doesn't at least occasionally love to be their own boss. That of course is no different on the homestead. Giving children dominion over a corner of the garden or responsibility for certain animals is a beautiful way to encourage budding little homesteaders. What might have felt like work or chores becomes something else altogether.

While some children will just want to plant their favorite veggies, having a themed garden can be loads of fun. For your very own pizza garden, plant basil, oregano, a variety of tomatoes, and any other veggies your family loves on a cheesy pizza. Our family loves corn and zucchini on summer pies! Children that love to drink their summer bounty can have a lot of fun with a tea garden. Lemon balm, mint, and chamomile are a great soothing place to start.

Make It Magical!

The longer we can allow our children to be children, the better for their development. While many children today are lost in worlds of screens, the homestead offers unparalleled opportunity to explore and create wild worlds within their imagination. Helping your children grow little hideaways for themselves or setting up stations outside that encourage open-ended play are much cheaper than a tablet and far more wholesome.

One of our favorite activities in early summer is to plant a sunflower house. Sunflowers are a wonderful gift to pollinators and can create a beautiful shady hideout for little ones. First gather

Credit: Julie Letowksi

Children can form special bonds with livestock.

a wide array of seeds for sunflowers that range in height, all the way from 1 foot to 12 feet when fully grown. Next, determine the desired size of your house and dig a circular or square trench in which to plant your sunflower seeds. Leave a two-foot section undug in your circle or rectangle for the entrance to the house. Plant your seeds in an even distribution and water, water, water! Even before the sunflowers bloom, children will have a wonderful spot all their own. In addition to sunflower houses, we grow cucumber and bean teepees for our son inside the actual garden. Doing so gives children their own little space to paint, pretend, or secretly chow down on the garden's bounty. Scarlet runner beans are excellent for a child's bean hideout as they are a prolific vining bean and they produce beautiful little red flowers.

Credit: Julie Letowski

A child raised on a homestead learns many skills that aren't always part of regular curriculums.

When I look back on my childhood, I have fond memories of being in my grandmother's suburban backyard with a couple of aluminum pie tins, mud, and orders from the family for delicious mud pie. The mud kitchen is a quintessential childhood experience and, on the homestead, a great way to occupy your older children while tackling that ever-growing to-do list. For parents of toddlers on up, the mud kitchen offers sometimes hours (yes, *hours*) of independent play. While you can find secondhand play kitchens online and in local thrift stores, it's incredibly easy to make them yourself. Our family scavenged through the barn and turned up an old wooden filing cabinet and nightstand that with a little paint and scrap wood were turned into a sunny yellow refrigerator and oven. From there we gave our son old dishes and kitchen tools, a shovel, and a bucket for water. We set up the mud kitchen on the side of our garden under the shade of the apple tree and have enjoyed it for years as a family since.

Educating Little Homesteaders

Homesteading families in general have the same options available to them for schooling as other families do: homeschool, public, and private. That said, many homesteads tend to be in more rural areas, and that does impact a family's choices significantly.

Homeschooling

Many families passionate about living the homestead life also find themselves passionate about homeschooling their children. With so many amazing opportunities for hands-on learning, this can be a great option, with the added benefit of a more flexible schedule. Let me tell you, it can be quite the predicament when a child needs to be picked up from school and the family milk cow is calving!

One of the greatest draws for homeschooling is how deep and wide the possibilities are. This makes it possible for every family to find their perfect style or curriculum, which is a huge asset to the children. The first place to start is with research on your state's homeschooling laws. Your children may need to be assessed by a certified teacher, build a portfolio over the course of the year to be submitted and looked over, or take an achievement test. You may also need to submit a letter to the state explaining your intent to homeschool. It really varies based on where you live. If you're just getting your footing as a homeschooler, within many rural communities, there are homeschooling co-ops and meet-ups for support and community, and the internet holds seemingly endless information on the subject. In recent years, homeschooling has moved more into the mainstream, giving interested families a great deal of resources from which to draw.

Public School

While it isn't always the case, many families transitioning to rural homestead living find themselves making a life within new communities. It can be an isolating experience. Public school is an incredibly easy and effective way to immerse yourself within the community and get to know your new neighbors. But here's the catch about public schools: many are great, some are less than, and a few are downright scary. Where your local school falls depends on a lot of factors. While you can look up test scores and ratings online, I have found a great way to really learn about schools is through conversation with parents. In our first years here, I stopped many a local parent at our farmers market booth to ask about their child's experience in the school. Are they happy? Is their child happy? Have you had any issues and how have those issues been handled? Parents care deeply about their children, and education is top of the list when it comes to their concerns. I promise, if you're curious, you'll only have to ask a few questions before the ball really gets rolling.

Private School

For some homesteading families, it may not be feasible to homeschool their children. Our family entered homesteading with an absolute certainty that we would homeschool. After all, we were, in large part, homesteading for our child. I couldn't imagine a world where our son would be off to school while lambs were being born or apples for winter applesauce were being collected. Cut to a few years later, we found ourselves with not only a super social only child but also off-farm jobs that were essential to the survival of our homestead. Homeschooling no longer felt like the right fit but neither did the local elementary school. Private schools can offer homesteading families the opportunity to give their children an alternative education experience when others are not feasible for whatever reason. With its emphasis on rhythms and the seasons, Waldorf education in particular is a great complement to the homesteading lifestyle.

School Choice

Talk about private school automatically feeds into talk about financing. Private schools are not cheap, even in rural communities, and homesteading, even when done really well, isn't known to be a gold mine. One of the greatest ways to make homesteading and private school work is to move to a town with school choices. For families who live in towns without local schools, sometimes school choice is an option, which means not only that you may choose where to send your child but that the town also pays a certain amount toward your child's education every year. The amount varies but can make a significant dent in private school tuition, and sometimes will even cover it in full. Looking for land within a town that offers school choice is an amazing way to open up options for your child's education.

The Right Way to Do It

The truth is, there is no right way to raise children on the homestead. What works for one family won't work for another. The same is true for parenting in general. The best advice I can give you is to always leave yourself open to whatever is working in the moment. This flexibility will save you a lot of heartache and stress that can come with trying to do two incredibly demanding jobs at once. Raising children on the homestead is deeply rewarding and important. Children raised in the great outdoors appreciate their surroundings and grow up to be stewards of the land, something

I believe to be essential nowadays. Through growing gardens, raising animals, and working alongside their family, they gain a deeper understanding of what it means to survive as a human on this Earth. Years into the process, I find myself regretting nothing. Good luck!

Further Resources

- Waldorf: Waldorf schools offer an alternative to traditional education that emphasizes the growth of the child. Their website offers resources and can also connect you to a Waldorf school near you. whywaldorfworks.org
- Montessori: Another alternative to traditional schooling, Montessori schools are all about hands on experiences. Visit their website to locate a school near you. livingmontessorinow.com
- Oak Meadow: A pioneer of homeschooling resources, Oak Meadow is a creative-based curriculum that can be used through high school grades. Oakmeadow.com
- HOME: While some of the resources on this site are Maine specific, or connect you to local events, others are universally helpful for those starting down the road of homeschooling. homeschoolersofmaine.org
- *The Well-Trained Mind* by Susan Wise-Bauer and Jessie Wise: One of the best primers on homeschooling your children, this book offers suggestions that are academically challenging while remaining home-based.

CHAPTER 9

Connectivity and Social Media

It is almost impossible to discuss "modern" anything these days without acknowledging the existence of social media. Social media, and social networks such as Instagram, provides you with a tool to reach potential customers and a way to make friends who understand your rural lifestyle, while at the same time it can also be a depressing and time-consuming outlet. Whether you are living on the farm or in the middle of a city, balancing the pros and cons of social media is a constant topic to consider.

Different Platforms

Before you set out to create a social media empire, think about what platforms you are interested in, and what tools will work best for your goals. If you want a lot of followers, for whatever reason, it is good to understand the platform inside and out before diving in.

Instagram

Image is the primary focus of Instagram. Farm life can often translate to beautiful images. You might not think that when you're covered in bits of hay or mud or manure, but essentially any animal can be Instagram gold, not to mention picturesque gardens or pastures, or vegetable produce and flowers fresh from the garden.

The key, then, to an effective Instagram feed is good pictures. Yes, a lot of other factors play into a successful profile, but the

Our goat Lucky is a social media hit, which helps attract people to our farm.

images are the foundation. Like the brick and mortar of a home, the rest really doesn't help if the pictures aren't there.

In this day and age, you do not need a fancy expensive camera to take excellent pictures. It certainly helps, but the newer iPhone models have very capable cameras. Choose dynamic photos, ones with your animals presenting interesting or funny expressions or performing some eye-catching activity, or take pictures of beautifully ripe produce. It's not enough to have a high-quality photo, it must have something that will command attention. Once you've got a good picture, take it to the next level using photo editing and playing with issues like the saturation, highlights, and shadows.

These “filters” are not cheating; every photographer uses them, and they can make the difference between an OK picture and an amazing one.

The easiest way to take a great photo? Get on your subject’s level. If you’re taking pictures of chickens, they’re going to look uninteresting from above. After all, that’s how people always see them! Kneel down, get on your animal’s level (or your plant’s, or your home’s), and snap the most interesting and dynamic angle of your subject. The difference between a photo taken from regular human-eye height versus right at the level of the animal is what catches interest.

Instagram offers a great community of farmers and homesteaders, and it can be a wonderful place to make friends and look at other people’s farms, as well as market your own. With over 700 million users as of April 2017, Instagram is one of the fastest-growing social media sites (via techcrunch.com, April 26, 2017). It continues to grow by leaps and bounds and is popular with all generations, making it one of the best ways to appeal to a large audience online.

Facebook

A behemoth of a network, Facebook boasts the most users of any social media site in the world. Its growth has slowed in recent years, and its active users skew older than some other social media sites, but slowing growth is still growth.

It makes sense to have a Facebook page for your farm or homestead if you’re trying to market any product: food, animal, fiber, lotion, or craft. In particular, Facebook’s Events section is a great way for interested people to find activities you might host at your farm, be it goat yoga, an open farm day, or a farmers market. Facebook will recommend Events to people who’ve attended similar activities, so they may find you even if they have not yet heard of your farm.

Instagram may be more focused on pretty pictures, but there will always be followers who prefer one network to the other, so if you have a successful Instagram, adding a Facebook page to the mix allows more people to find out about you.

For social networking, making friends, and finding other farmers, Facebook can be helpful, but you don't need to keep a homestead page in addition to your personal one. Facebook as a social media network in general has its pros and cons, which are largely personal preference, and you may find you enjoy its newsfeed or that you want less of your friends' opinions about your life.

Twitter

Twitter is an interesting network and a bit of an enigma. I personally do not see it serving much purpose for most homesteaders, certainly not as a business outlet. If you really want to focus on networking, Twitter does connect you with all kinds of users. It is an odd platform because it is text-based; the number of characters you can use is limited. So you cannot share images of daily farm life the way you can on Instagram, but you also can't wax lyrical about farm living. While there is a place for Twitter, I hesitate to say it is on the homestead.

Pinterest

Despite lagging behind Instagram and Facebook in terms of active monthly users, Pinterest can be a useful network for a homesteader. It may not be the primary place you connect with customers or fellow farmers, but it is a place for inspiration.

Pinterest's mission statement is "to help people discover and do what they love," and they do this by providing you with a platform to save images and articles from every corner of the internet. They allow you to organize on "boards," like digital bulletin boards, which can be focused on any category you prefer (or you can just keep a big mishmash, Pinterest lets you decide). The beauty of Pinterest is that you can make it what you want.

Homesteaders always have ongoing projects. Pinterest gives you a space to tack up digital ideas for your barn improvements or your dream house plans. Say you are always busy cooking and canning your fresh produce, Pinterest gives you a place to categorize that obscure herbal recipe you found alongside foraging tips from your favorite blogger. Even if you don't use Pinterest in any out-facing way at all (not trying to gain followers or engage friends), it is still a useful site.

You can create boards that are gift guides, pin your latest blog posts, or share your garden pictures on Pinterest to attract followers. The downside of Pinterest is that many people repin items without credit, so if you are posting an image be sure to "watermark" pictures (put your logo somewhere on the image when editing it) so people can get back to the source.

Blogs and Websites

Another challenging social media form to maintain are blogs. A personal blog also gives you the most space to express yourself to your followers. There are dedicated blog hosting websites such as WordPress and Blogger, as well as sites that can host both a blog and an ecommerce site, such as Wix and Squarespace. What platform you use is largely based on personal preference. Because domains cost money, I recommend going with a site that hosts ecommerce if you think you'll be selling anything in the future.

A blog means time and effort, and if you are not updating it regularly, your site quickly starts to look abandoned. Don't begin a blog if you know you won't have time to keep up with it, and be realistic about how much time you have versus how much time writing will take you. It's best to create a schedule for yourself, including a list of topics as well as a calendar of times that you plan to post. You will want to update your blog regularly and block out time for yourself to get the pieces written and published, perhaps every week or every other week. Fewer than one post a month is not enough, although more than one post a week can be excessive.

A blog serves a different purpose than many other social media connections. It isn't just about the pictures or the events, it's a place to tell your farm's full story or share your personal philosophies and impressions. If you're moving out to the land and you want to tell your city family back home what it's like, a blog might be a good choice for you. If you enjoy writing, it's a creative outlet, and it can complement other social media nicely because there's no text limit or restrictions about what you can and cannot post.

Blogs are also the easiest way to get homesteading information online. Many of the tips and tricks I've picked up over the years have come from homesteaders' blogs, where anecdotal information is passed from resourceful farmer to curious adventurer. Blog information, of course, isn't necessarily scientifically researched so it's important to back up what you learn online with resources from bookstores or other reliable sources.

Other Networks

Plenty of other social media sites are out there, and more emerging every day. One steadfast way to reach followers is via YouTube, where plenty of "vloggers" host their video blogs and share tips, give demonstrations, or present new ideas in a visual format. YouTube is a place to both find and share information that might be otherwise hard to describe.

Snapchat and other newer social media sites are very popular with the younger generation, and it's good to understand these networks so that you won't be left behind as they gain popularity. However, at this time, there isn't a lot of marketing reason for a homesteader to be on Snapchat; it is just a fun way to connect with friends who happen to be farmers.

Marketing Use

While not every homesteader has a product that they are looking to sell, most will at least want to supplement income with farm-grown produce or homemade crafts. If you are marketing anything from fresh produce to hand-knit sweaters, social media is a great way to get people far and wide interested in your brand.

For farmers growing produce, making milk or cheese, or raising animals for meat, social media provides a glimpse behind the curtain for their customers. It allows them to feel like they're on the farm without having to get their boots dirty, and it helps them to better understand what goes into the production of their food. It might seem like everyday stuff to you, but to your customers, your life can be fascinating. Getting in touch with what you do and how you do it cultivates an attachment for the viewer to your farm, and your products.

If you make something not strictly farm produced, but related to your homestead lifestyle, this still applies. Craftsmen of all kinds use social media to their advantage, be they woodworkers, chefs, makers of textiles, or even writers. Social media broadens your audience, and that is useful for your business.

The main complaint I hear about social media is that it presents a highly curated, picture-perfect version of life to the outside world. If you are marketing something, that is not necessarily a bad thing. A few "#reallife" posts that show a splash of mud on your face or the mention of long days turning soil is still relatable. If you have a separate personal account on social media, that is the space where you can rant about your ex-boyfriend or share your detailed opinions on a political situation. Your business social media is different. It should appeal to the widest audience, especially if you keep it clean and happy. It is not deceptive: those clean, happy events are happening in your life, just in addition to all of the messy ones.

The obvious exception to this is if your business is political or activist in nature, in which case post all of the long-winded rants you want!

You'll have to learn the hashtag game if you are using any kind of social media, Instagram and Twitter especially. A few fun hashtags that are unique to you will help you find pictures later on and appeal to witty millennials, but the hashtags that get you followers are the ones that viewers search for every day. For example, #goatsofinstagram: there are people who will sit down with a glass of wine in the evening and just review what was posted under that hashtag in a single day. So, use it! They will find you. The best way to figure out what hashtags will work for your business is to look at pages similar to yours that are succeeding and use the tags they use.

Instagram is often the best forum for this kind of connection, where a picture is worth a thousand words and a quick caption is all you need to spend time on. However, a blog, if you have time to keep up with it, is the most expansive way to tell the stories of your farm, stories that will strengthen that relationship with your customers. Again, remember that to them your life can be a fascinating new world, even if you feel like it's just the same old chores you do every day. Facebook is also helpful, specifically to promote events at your farm and to appeal to a slightly older audience.

The biggest downside to using any kind of social media as a marketing tool is its total lack of locality. While Facebook tries to target people geographically near you occasionally, most social media is all about connecting to people from around the world. If you've got a product that can ship, that is fantastic and will broaden your reach overnight. But if you are selling fresh produce or trying to get people out for a local farmers market, you can have thousands of Instagram followers and only a handful are able to turn up for the event.

Building a Community

For the lonely homesteader, building community may be the biggest benefit of a social network. Instagram is the place to discover like-minded people who might be facing the same challenges you are.

As you start using a social media platform, it can be wholly intimidating. I kept an Instagram account for a full year before I used it, unsure exactly how to proceed. Opening an account but not using it at first is a good plan of attack. You can start browsing other users' profiles and figure out what is going to work for you, and then start posting.

Most of the people I have connected with via social media have been very friendly, encouraging, and helpful. It's important to remember that really popular accounts probably get several messages a day, and even more photo comments. Instagram isn't the best at updating you if you have a ton of such comments, so it's easy to miss one or two. Facebook is much better at this, but either way, if you get ignored by an account with thousands of followers, don't take it personally. Keep on reaching out if you have questions, and most people will be happy to help you out.

Knowledge

I utilize Instagram and Facebook as knowledge resources all the time. An old-fashioned Google search can turn up a blog post on a subject I am curious about, still, connecting with someone via Instagram is rewarding because it gives that extra personal touch. If you take the time to read the captions of people's posts in addition to looking at their pictures, you can find surprising bits of knowledge as people explain how they're coping with day-to-day problems around the farm. It is definitely a resource, and even more so if you start connecting with other Instagram farmers and share ideas.

Instagram and Facebook are also where people share updates to their blog, so if you want to make sure you're connected to someone's writing, follow them on Facebook or subscribe to a blog newsletter. And do not be afraid to ask! If you see someone with a farm like yours, maybe keeping the same type of livestock guardian dog while you're trying to train yours, send them a message and ask for tips. Most people are happy to help out. A message is more effective in connecting to someone as comments can get lost on a busy page.

Friendships

Friendships can be the most valuable aspect of social media for the rural homesteader. Social media provides you with quick easy outlets to show folks back home what you are doing. Yes, personal letters and phone calls are intimate ways to reach faraway friends and family, but sometimes a quick Facebook, Instagram, or blog post will reach a wider audience and succinctly express more of your day-to-day life.

In fact, many homesteader blogs I know started as a way to share experiences with folks back home, and grew into reaching other homesteaders as well. Instagram and Facebook also offer an instant way to get pictures to your friends and family, pictures that express what you might not be able to explain in a phone conversation.

Bonding with other homesteaders is a useful aspect of social media. It connects you to other people in rural settings who you would not run into otherwise. It brings people together who may be struggling with the similar issues and applauds your private victories. In this way, social media can be downright invaluable.

It can take as much time to build friendships online as it does in real life. You not should expect to be able to sign up, post a photo or two, and immediately find a slew of like-minded people. It is often best to start by tracking down local farms that you are aware of, or check and see if the nearest farmers market or country store has a profile. You can also try finding other names you might be familiar with, like hatcheries and seed catalogs or even authors whose books you enjoy. Follow and "like" what you are interested in, and people will follow you back.

If you're ever at a loss and overwhelmed by Instagram specifically, there are a couple of tips to connecting to people similar to you. The Search icon at the bottom of the Instagram app takes you to images selected just for you, based on who you follow and what you like. It's another way to discover new profiles. If you are totally new to the app, it may not be accustomed to your tastes yet. Instead, if you click on the heart at the bottom of the app, it shows you who has recently liked your photos. Swipe right on that screen, and you'll see what images the people you follow have recently liked.

While I recognize the pitfalls of social media in general, it has been very good to me. I have not only been fortunate enough to gain many followers interested in our farm's daily workings and our animals, I've also made some invaluable friends there. Most notably I have found multiple friends with farms nearby that I would not know were there otherwise. These people have ended up being good friends and folks who can come over and help out with the animals or who we enjoy spending time with socially. They are real-life friends that I might not have found if I had simply hoped to meet accidentally by bumping into them in the feedstore.

The people I've met through social media include the owners of the buck goat that we have used and plan to use to breed our goats, the guest author of chapter eight of this book, and the photographer for many of the images in this book who has also become a good friend. There are the proprietors of small farms and apothecaries where I now make regular orders. Whether close or faraway, I've found all of these real-life friends to be enthusiastic supporters, quick to comment with tips and suggestions when something is wrong, who likewise offer excitement and praise for good news. My community of friends online aren't just supportive, they're the one group I know will understand the various odd challenges of homestead living.

Trolls

In spite of these praises of social media and the people you can connect with there, there is a downside. There are unpleasant folks who make a habit of posting ridiculing comments, people who spend their time searching out specific pages and posts that they disagree with. The best way to deal with these "trolls" is to not engage. Most of these them are itching for a reaction; the argument is what excites them, and they don't consider the person behind an account as a real human with feelings. If you are going to respond to any negative comments on your feed, do so in a well thought-out, calm, and polite manner. A shouting match will just descend into name-calling. You might think this is only going on privately between you and your antagonist, but everyone else who follows you is seeing it as well. Trolls might even be getting notifications every time you post, and you stand the chance of losing other followers who just don't want that kind of drama in their feed.

It is often best to not respond at all. If you must, respond politely to clarify the situation without getting into defensive details. The most common source of online drama I have found has been vegans versus meat eaters. I follow a number of profiles of people who raise meat animals for a living; their main mission is to provide animals with a happy, healthy life before they become food. They are regularly trolled by tribes of PETA-supporters. Each time, a back-and-forth argument develops that ends with people blocked on both sides. I don't have a solution for the arguments except to try to understand where each side is coming from.

Pitfalls

Social media has other drawbacks besides the possibility of getting into a debate about ethical butchering. Across society there has been plenty of commentary linking social media to clinical depression. Looking at picture-perfect posts can be discouraging if you are constantly comparing your own circumstances with someone else's. Pangs of envy can arise when your friends post baby goats or new chicks or something else glowing happy and looking carefree. This can become a problem if you are prone to discouragement.

Everyone handles this sort of social pressure differently. For me, I remember that social media *is* about pretty pictures. Remember, these other farms no doubt have as many trials and tribulations, sometimes far more serious setbacks, as everyone does. Posting cute baby animal pictures every day doesn't mean that their farm doesn't have its troubles. The platform is set up that way, and if you're using it, you should understand you're going to see the best and most positive images of people's lives.

If you find yourself getting depressed over others' social media lives, or ending up with extra animals you don't need just to keep up with the digital Joneses, maybe you should back out of social media. It isn't a necessity to homesteading. If you don't enjoy it, don't do it, and your farm will be no less for it.

Personal Enjoyment

For some, social media is about how many followers you have. Some of the followers can translate into dollars; some can translate into friendships. But if you do not enjoy social media, it isn't serving its purpose. Apart from the possible emotional angst of seeing other farmers' successes, there is also the continual push for more followers. Instagram and Facebook can become quite competitive, and you may find yourself checking your numbers every day and comparing them against those of your friends and online rivals. Like high school, it becomes all about popularity.

And if that is the case, you'll also be spending more and more time taking pictures. I spend at least a couple of hours a week taking pictures, and many more hours editing and generating captions. That is time totally separate from the maintenance of my blog or any other writing projects that I am working on.

The golden rule is: does it make you happy? If you love taking pictures (which I do, let's face it: I'd be outside clicking away regardless of the outlet I had for the pictures), and you enjoy the people you've found online, then it is a success and it's right for you. If it stresses you out, you feel guilty on days that you do not post, and you are obsessing over everyone else's opinion of your farm, it's not worth it. You aren't homesteading to be the coolest kid in the forest, you're doing it as a lifestyle choice, and social media should be a pleasant and enjoyable addition to that, not a burden or an end goal.

Off-grid Living

I've mentioned a few times already that I do not believe that social media can be achieved if you're living off the grid. While social media isn't technically part of the infrastructure of this country, generally speaking the goal of off-grid living is to disengage. By contrast, social media seeks out engagement. I personally enjoy its use, but I do not claim to be an off-grid homesteader. I'm sure some folks would disagree with me, but it's my opinion that off-grid living means cutting out those kinds of connections to the outside world.

Questions Before You Leap

- Do you have or use social media? If yes, what platforms?
- Are you confident with your photography skills?
- Do you enjoy writing?
- Do you want to use social media primarily to promote your farm and produce?
- Do you want to use social media primarily to connect with people, make friends, and as a hobby?
- Do you have the time to actively maintain a social media presence?

CHAPTER 10

Rewards of Rural Living

There are many unromantic realities to rural living. Some, like feeding and tending animals and gardens, you may have come to expect. Others, like the remoteness and the bothersome neighbors, may be a perturbing surprise. But can I say without hesitation that all such hassles and annoyances are worth it for the immense pleasure and satisfaction of homesteading life. The romance of rural life is a reality, and the hardships far outweigh the joys.

Moving with the Seasons

If you think that you notice the subtle shifting of the seasons in an urban or suburban setting, you will be amazed at how each week transforms into a new, intriguing world on the homestead. Seasons do not have the blunt beginnings and endings of the calendar dates, and you will be hard-pressed to divide them into four concrete seasons.

Spring shows up as early as late February when, on our little holding, the ganders start getting feisty, the slant of the light is noticeably different, pussy willows' tender catkins are budding, and sap starts rising. By mid-March, it is full-on syrup time. We discover that more and more jobs can get done in a single day as the daylight creeps in for longer periods with each passing week. The bitter winter winds of Maine turn from cold and dry to wet

Spring is marked by increasing egg production, even before the snow has melted.

and, well, still cold. Around our area, a storm scan now signals a warm front moving in, instead of a cold one.

After the pussy willows, come the snowdrops, delicate white flowers show above mossy knolls. Then early crocus, and finally after many weeks, the tulips and the daffodils are late-comers to the spring party. What seems like endless acres of mud around every outbuilding in March seeps into the earth and transforms itself into pale green shading every tree branch, and soon that shady dusting turns into sticky new leaves on birches and maples.

The headiness of spring can seem magical after a long winter, but change remains constant and subtle throughout each season. In summer, every week is marked by a change for the gardener, rotating from peas and radishes to zucchini and cucumbers to tomatoes and eggplants. The forager also finds a changing harvest every few weeks, from dandelions and elderberry flowers to milk-weed buds and daylilies to elderberry berries and fresh acorns.

Autumn, too, is not an overnight transition. Each species of tree loses its leaves at different times, showing off different shades of yellow, orange, red, and brown. Finally, the only "leaves" left are the needles on the evergreens, when the gray skies and high winds of November chill the air.

These changes, however subtle, are punctuated by the required tasks: digging garden beds, or raking leaves for mulch or compost, or clearing snow. Animal coats grow in thick and fuzzy as the days get short, then shed in thick clumps, carried off by birds for nests, sometimes as early as February. Rivers swell, shrink, dry up to a trickle, swell again, and then burst their banks as ice blocks and melting snow flow downstream. All of this, depending on your location, can directly impact your homestead and be a blessing or a burden. Our farm's spring-fed well runs dry every August, despite our most careful water use. But it's back and bubbling by October. Areas flood in spring, looking like they'll never be dry again, only to return to clear pastures within a week or two. And of course, roads become icy and then muddy, almost impenetrable, but that also passes.

Elderflowers, which can be foraged in Maine by June, make delicious cordials and fritters.

Privacy

If I were to sum up why I wanted to move to a rural location in a single reason (leaving out my desire to expand my flock of geese and generally have more animals), I would say I was looking for privacy.

People gossip about their neighbors everywhere, it must be said. But at least they cannot look in your windows. In our former suburban home, every move felt as if it were taking place in a fishbowl, and I felt on display, under the watchful scrutiny of the nearby neighbors.

When we started our search for a new life of homesteading, we looked at places at the end of long dirt roads, miles from anywhere. I wanted property where I could go naked if I chose and

Relaxing is an important part of homesteading happily, and knowing no one will be dropping by to say hello helps you mentally unwind.

not scandalize anyone, a space where the raucous honking of our gaggle of geese would not bother anyone, and where we could target practice without neighbors calling the police. We definitely found it, even though the road that runs through our ninety-plus acres is technically public, it is still off the beaten track, and it fits the bill for "privacy."

For some, privacy is an adjustment, but for us it is not only appealing but a requirement. When we first found the property we purchased, we spent several hours there waiting to see how many cars passed by. It averaged out to two or three a day, and that included the daily mail delivery.

In many areas around the country, you have to worry about city ordinances before getting a flock of chickens. In rural Maine, that is not the case—I would guess that there are more chickens in our small town than people. In fact, there's a flock at the post office that often block my entrance to the building clucking for treats. Gossip is as rampant as it is anywhere, but dropping out is far easier. I can go weeks at a time without seeing another soul, if I choose to. At the end of the day, we cannot see lights from anyone else's homes. Step outside at night, or first thing in the morning to feed the goats, and the silence wraps around you.

Even if your property itself is not large, the country offers plenty of space to stretch your legs. Driving across the country, you may see city after suburb, farm after factory, but in fact under 10% of America's land is fully developed. Almost half of the country is totally unoccupied. So it is surprisingly easy to step out the door for a walk, and spend hours immersed in the surrounding natural wildness.

Self-confidence

It might be the lack of worrying about keeping up with the next-door neighbors or the ability to spend time working alone or along with my partner as we clear brush, clean stalls, and shingle the barn, or it might be the pride of gaining new skills like driving a tractor and the satisfaction of assisting with the birth of goat kids, but my confidence level has developed remarkably since moving to the country.

I used to spend way too much time worrying over my public appearance, especially before heading out to the office every weekday morning. Now I sometimes go days without showering or even changing into clean jeans. I don't mean to imply that my personal habits have deteriorated into a vagrant's life, it is just that on the farm I do not find it necessary to put on special duds for an audience consisting of a gaggle of geese and a herd of goats. If I am going out, I don't see the need to present myself as anything other than what I am: a farmer.

Not that I condone being unhygienic. The main reason I am no longer self-conscious about my attire and fashion accessories is simply that I don't have the time. After a morning of tending goats, trimming hooves and worming them, some wormer-drool

The outdoor shower keeps us clean and provides stunning views when bathing.

is left staining a portion of my sweatshirt. Then there's collecting eggs and cleaning out the nesting boxes, which routinely leaves decorative chick droppings on my already mud-encrusted rubber boots. Oh, then I realize that we have to go get some more hay and shavings for bedding down the expectant goat's stall; she is due to kid any day now. There isn't time or inclination to do much more than wash my hands and face and promptly get to the feedstore in town. Days like this are my new "normal." Regular days on the homestead are filled with chores that don't require fancy wardrobes and worrying about what the guy at the feedstore thinks of my hairstyle.

Country living is a crash course in taking care of the farm, the animals, the garden, the fields, the fences, the harvest. All this self-reliant business is a boost to one's own self-confidence. If something needs to be done, then I know that either I will do it, or my partner will, or we will do it together. We are responsible for what we started here, and it is only when I stop to think about it that I realize what we accomplish every single day, how much we've managed to do since we got here, and what more we are truly capable of achieving in the days to come.

Skills that we've talked about in this book, like sharpening an axe or getting a rototiller running, are skills that require muscle you didn't know you had. And then there's the skills, like delivering or bottle feeding a baby goat, that require persistence and patience, a whole different side of the brain. If nothing else, you become well-versed in the skill of adaptability.

I think most homesteaders and farmers would agree, life on the farm is the best workout program around. There's nothing like hauling and lifting fifty-pound bags of feed or tossing hundreds of bales of hay around to get your body fit and keep it that way.

The Stars and the Silence

The open space of natural land in the country is nothing compared to the vastness of open space in the sky above, especially on a cloudless night. Constellations stretch twinkling across the jet-black sky. The moon marks time with each phase noticeably different, so that by the time of a clear full moon, curtains might be as necessary as they are in a city in order to get a good night's sleep. Otherwise, without street lights and noisy night-time traffic, the silent spell of nature is as comforting as a lullaby.

Even in the daytime, the sky stretches all around the homestead. Each small seasonal shift indicates a change in local wildlife, specifically birdsong. In the summertime, no matter how early you rise, you won't beat the morning's cacophony of eager songbirds, all vying for attention. Spring starts with just a few trills at dawn, the titmouse and a few finches just arriving. During the day, you may see a flock of robins as early as February, and by the time the grass is greening, there will be splashes of color all over your fields.

Peepers are a reliable soundtrack to spring in our area, sealing the fact that winter is over. Peepers are small frogs that emerge before other amphibians in late March or early April, a chorus of peepers might begin with just a few lonely, high-pitched mumbles and crescendo into a course of a thousand chirping, froggy voices, all heralding the coming spring.

Birds are often the reason why many farmers hay when they do: either before the birds of the fields (namely field swallows and bob-o-links) have nested, or once those chicks have safely fledged. The appearance of barn swallows is a guarantee that summer is on the way. Birds that I see every day of the year are clever crows, often silent, who pillage any food our geese drop from their beaks.

In wintertime, however, the absence of birdsong is equally notable. Many mornings, I've come out to feed the goats and wondered what on Earth has happened that could cause such total silence. The lack of sound rings in my ears. Once the sun starts to creep above the horizon and the geese start honking, the dog barks. The silence of a country night can be unnerving. You do have to embrace a certain confidence and peace with yourself to not feel unnerved when that utter silence is broken only by the distant hoot of a night owl or the yapping of a pack of coyotes.

To cultivate confidence in the silence, learn to identify the sounds of your land. If you know what is out of the ordinary, it is much easier to relax when the faint calls of the forest ring out in the night. In fact, learning the various calls of the wildlife around you can be a fun activity for summer evenings, and soon you'll know all of them by heart. That also means that when something out of the ordinary sounds, especially if it's near your chicken coop, you can jump to action quickly.

Healthier, and Better for Kids

Health is not the primary reason that every homesteader points out as their inspiration for going rural, yet that is a call for many, and it ties into the story of self-sufficiency. Those who have a dispute with modern agricultural practices or the pharmaceutical industry are homesteading to preserve their health.

Both city and country living have their health benefits and risks. Many rural statistics are skewed by the fact that the rural health care infrastructure is often lacking. Emergency rooms can be far away, and many people function without insurance. However, the fact that country living is generally healthier, with growing organic food, breathing clean unpolluted air, and getting plenty of exercise, is pretty inarguable. With cleaner air and less pollution, country people are less likely to develop conditions like asthma and allergies.

Rural residents are also far less likely to suffer a variety of mental disorders. Gizmodo.com reported in 2014 that the rate of mood disorders among city dwellers is 21% higher than with their country equivalent, and risk of anxiety disorders can run 39% higher. Cities also can affect people's internal clocks, with air pollution and late nights wreaking havoc on the natural rhythms of the body. While this stressor doesn't necessarily have an immediate impact, you may find yourself with a higher risk of depression, inflammatory diseases, and cancer in the long term.

It is interesting to look at the concrete health consequences of rural life. There is a gap in life expectancy, with city dwellers living longer; there are more fatal car crashes and the suicide rate in the country is much higher than in the city. Crime is even higher in the country than the city, specifically the risk of violent crime.

However, these numbers are misleading. They are skewed partially by the health care infrastructure, but also by the income levels and lack of education in rural areas. For a family or individ-

Homesteading life brings deeper bonds with the land and animals.

ual who can afford to eat a healthy diet or grow their own food, the country provides clean air and space to exercise. Smoking rates, and therefore lung cancer, are much higher in the country, as are drinking and driving rates. However, it could be argued that these rates are more education based and less a consequence of the country being a tougher place to live.

Country living can certainly provide relief from the day-to-day rat race of the city. There is no arguing that there is less pollution and fewer mental stresses. Oftentimes the problems you face and need to conquer are situations that put you in control—ones you can figure out for yourself. Country life is empowering, and that feeling translates to a healthier life.

If you have children, it's also good to take into account that they will have a safe space to play outside, learn various life skills from homestead living, and understand nature and where our food comes from in a way city kids don't comprehend. Schools in rural areas can be poor, but that setback can be overcome. A teenager may complain about the lack of social life or events to attend in the country, but this too can be resolved if parents are willing to seek out other teens in the community and sponsor teen activities.

I have known homesteaders from all walks of life. Some grew up in an environment with a garden or animals, some grew up in the city. There are men, there are women, people of all gender identities and ethnic backgrounds, women who've worked as models, and men who came from Wall Street who have found satisfaction and a home doing the job of homesteading. "Homesteading" may be hard to define, but the people who choose to lead this lifestyle are even harder to put into one archetype or another. For the most part, they share some kind of disillusionment with society, whether it is big agriculture, big pharma, government surveillance, or other particular issues. Yet all of these people have to perform the unexpected at some point on their homestead. The former supermodel figured out how to set up a tractor-driven generator to power the lights in a storm outage.

The burly ex-Marine is tenderly assisting in the birth of a calf. Although these jobs might take people completely out of their comfort zone, they are the very ones that bring out the faith in your own competence. Experiences like these are almost always deeds and actions that need to be done in the moment. Stepping up to the plate, homesteaders realize all of the amazing things that they can accomplish with a little bit of urgency or focus.

Index

About the Author

KIRSTEN LIE-NIELSEN grew up on a farm and has been raising geese and enjoying the quirky personalities and practicality of these beautiful birds for most of her life. Always intrigued by self-sufficiency and working with her hands, Kirsten and her partner are restoring a 200-year old farm in Liberty, Maine, where they raise animals and grow vegetables and native medicinal herbs. Kirsten writes about her experiences and the lessons of life on the farm for publications such as *Grit* and *Mother Earth News*, *Backyard Poultry*, and Hobbyfarms.com. She also blogs about the good life at HostileValleyLiving.com.

More Resources for Homesteaders

The Modern Homesteader's Guide to Keeping Geese
KIRSTEN LIE-NIELSEN
7.5 × 9" / 144 pages
8 page color section
US/Can $19.99
PB ISBN 978-0-86571-861-6
EBOOK ISBN 978-1-55092-654-5

Raising Rabbits for Meat
ERIC and CALLENE RAPP
7.5 × 9" / 208 pages
US/Can $24.99
PB ISBN 978-0-86571-889-0
EBOOK ISBN 978-1-55092-682-8

Raising Goats Naturally, 2nd Revised & Expanded Edition
The Complete Guide to Milk, Meat, and More
DEBORAH NIEMANN
7.5 × 9" / 352 pages
US/Can $29.99
PB ISBN 978-0-86571-847-0
EBOOK ISBN 978-1-55092-642-2

The Frugal Homesteader
Living the Good Life of Less
JOHN MOODY
Foreword by Joel Salatin
7.5 × 9" / 224 pages
US/Can $24.99
PB ISBN 978-0-86571-893-7
EBOOK ISBN 978-1-55092-686-6

www.newsociety.com

ABOUT NEW SOCIETY PUBLISHERS

New Society Publishers is an activist, solutions-oriented publisher focused on publishing books for a world of change. Our books offer tips, tools, and insights from leading experts in sustainable building, homesteading, climate change, environment, conscientious commerce, renewable energy, and more—positive solutions for troubled times.

DON'T EAT THIS BOOK (but you could)

We're proud to hold to the highest environmental and social standards of any publisher in North America. This is why some of our books might cost a little more. We think it's worth it!

- We print all our books in North America, never overseas
- All our books are printed on 100% **post-consumer recycled paper**, processed chlorine-free, with low-VOC vegetable-based inks (since 2002)
- Our corporate structure is an innovative employee shareholder agreement, so we're one-third employee-owned (since 2015)
- We're carbon-neutral (since 2006)
- We're certified as a B Corporation (since 2016)

At New Society Publishers, we care deeply about *what* we publish—but also about *how* we do business.

Download our catalog at https://newsociety.com/Our-Catalog or for a printed copy please email info@newsocietypub.com or call 1-800-567-6772 ext 111.

New Society Publishers

ENVIRONMENTAL BENEFITS STATEMENT

For every 5,000 books printed, New Society saves the following resources:[1]

18	Trees
1,617	Pounds of Solid Waste
1,779	Gallons of Water
2,321	Kilowatt Hours of Electricity
2,939	Pounds of Greenhouse Gases
13	Pounds of HAPs, VOCs, and AOX Combined
4	Cubic Yards of Landfill Space

[1] Environmental benefits are calculated based on research done by the Environmental Defense Fund and other members of the Paper Task Force who study the environmental impacts of the paper industry.